Smart Computational Intelligence in Biomedical and Health Informatics

T0309293

Computational Intelligence Techniques
Series Editor: Vishal Jain

The objective of this series is to provide researchers a platform to present state of the art innovations, research, and design and implement methodological and algorithmic solutions to data processing problems, designing and analyzing evolving trends in health informatics and computer-aided diagnosis. This series provides support and aid to researchers involved in designing decision support systems that will permit societal acceptance of ambient intelligence. The overall goal of this series is to present the latest snapshot of ongoing research as well as to shed further light on future directions in this space. The series presents novel technical studies as well as position and vision papers comprising hypothetical/speculative scenarios. The book series seeks to compile all aspects of computational intelligence techniques from fundamental principles to current advanced concepts. For this series, we invite researchers, academicians, and professionals to contribute, expressing their ideas and research in the application of intelligent techniques to the field of engineering in handbook, reference, or monograph volumes.

Computational Intelligence Techniques and Their Applications to Software Engineering Problems
Ankita Bansal, Abha Jain, Sarika Jain, Vishal Jain, Ankur Choudhary

Smart Computational Intelligence in Biomedical and Health Informatics
Amit Kumar Manocha, Mandeep Singh, Shruti Jain, Vishal Jain

For more information about this series, please visit: https://www.routledge.com/ Computational-Intelligence-Techniques/book-series/CIT

Smart Computational Intelligence in Biomedical and Health Informatics

Edited by

Amit Kumar Manocha, Mandeep Singh,
Shruti Jain, and Vishal Jain

CRC Press
Taylor & Francis Group
Boca Raton London New York

CRC Press is an imprint of the
Taylor & Francis Group, an **informa** business

First edition published 2022
by CRC Press
6000 Broken Sound Parkway NW, Suite 300, Boca Raton, FL 33487-2742

and by CRC Press
2 Park Square, Milton Park, Abingdon, Oxon, OX14 4RN

© 2022 selection and editorial matter, Amit Kumar Manocha, Mandeep Singh, Shruti Jain, and Vishal Jain; individual chapters, the contributors

First edition published by CRC Press 2022

CRC Press is an imprint of Taylor & Francis Group, LLC

Reasonable efforts have been made to publish reliable data and information, but the author and publisher cannot assume responsibility for the validity of all materials or the consequences of their use. The authors and publishers have attempted to trace the copyright holders of all material reproduced in this publication and apologize to copyright holders if permission to publish in this form has not been obtained. If any copyright material has not been acknowledged please write and let us know so we may rectify in any future reprint.

Except as permitted under U.S. Copyright Law, no part of this book may be reprinted, reproduced, transmitted, or utilized in any form by any electronic, mechanical, or other means, now known or hereafter invented, including photocopying, microfilming, and recording, or in any information storage or retrieval system, without written permission from the publishers.

For permission to photocopy or use material electronically from this work, access www.copyright.com or contact the Copyright Clearance Center, Inc. (CCC), 222 Rosewood Drive, Danvers, MA 01923, 978-750-8400. For works that are not available on CCC please contact mpkbookspermissions@tandf.co.uk

Trademark notice: Product or corporate names may be trademarks or registered trademarks and are used only for identification and explanation without intent to infringe.

Library of Congress Cataloging-in-Publication Data
A catalog record has been requested for this book

ISBN: 9780367624125 (hbk)
ISBN: 9780367624149 (pbk)
ISBN: 9781003109327 (ebk)

Typeset in Times
by KnowledgeWorks Global Ltd.

Contents

Figures

Tables

Preface

Health informatics involves multidisciplinary domains to extract information and knowledge from physiological data to use in decision making for improved human health through the effective use of recently developed technologies and algorithms. The aim is to provide a cross-disciplinary forum to share information on research, simulations and modeling, measurement and control, analysis, information extraction, and monitoring of physiological data in clinical medicine and the biological sciences. Emphasis is placed on contributions dealing with the practical, applications-led research on the use of methods and devices in clinical diagnosis, disease prevention, patient monitoring, and management. Health informatics is closely related to artificial intelligence where heuristic as well as metaheuristic algorithms are designed to provide better and optimized solutions in reasonable amounts of time. These algorithms have been successfully applied to different application domains in biomedical, bioinformatics, and biological sciences. The practice of recent biomedical research requires sophisticated information technologies to manage patient information, and plan for diagnostics, prognostics, procedures, interpretation, and investigations. This provides a conceptual framework and practical inspiration for the quickly growing and promising engineering and scientific disciplines of computer science, decision science, information science, cognitive science, and biomedicine. The objective of this book is to provide the researchers a platform to present state-of-the-art innovations, research, design, and implement methodological and algorithmic solutions to data processing problems by designing and analyzing evolving trends in health informatics and computer-aided diagnosis. This book will provide support and aid to the researchers involved in designing decision support systems that will permit the societal acceptance of ambient intelligence. The overall goal of this book is to present the latest snapshot of the ongoing research as well as shed further light on future directions in this space. This book presents novel technical studies as well as position and vision papers comprising hypothetical/speculative scenarios.

Amit Kumar Manocha
Mandeep Singh
Shruti Jain
Vishal Jain

Acknowledgments

We want to extend our gratitude to all the chapter authors for their sincere and timely support to make this book a grand success. We are equally thankful to all CRC Press executive board members for their kind approval and granted permission for us as editors of this book. We want to extend our sincere thanks to Dr. Gagandeep Singh and Mr. Lakshay Gaba at CRC Press for their valuable suggestions and encouragement throughout the project.

It is with immense pleasure that we express our thankfulness to our colleagues for their support, love, and motivation in all our efforts during this project. We are grateful to all the reviewers for their timely review and consent, which helped us improve the quality of this book.

We may have inadvertently left out many others, and we sincerely thank all of them for their help.

Amit Kumar Manocha
Mandeep Singh
Shruti Jain
Vishal Jain

Editors

Dr. Amit Kumar Manocha is presently working as an Associate Professor in Electrical Engineering at Maharaja Ranjit Singh Punjab Technical University, in Bathinda, India. Dr. Manocha obtained his Ph.D. in 2015, M.E. in 2006 and B.Tech in 2004. He is the author of more than 50 research papers in refereed journals and international and national conferences. Dr. Manocha successfully organized five international conferences in the capacity of conference chair, convener, and editor of conference proceedings and more than 25 workshops and seminars. He participated in many international conferences as an advisory dommittee member, session chair, and member of technical committees in international conferences. He is a member on editorial boards for many international journals. His area of research includes Biomedical Instrumentation, Remote Monitoring and Control Systems. He has guided more than 10 Master's degree and five Ph.D. candidates. He has been granted/published 03 Patents and Rs. 36 Lacs from the Department of Science and Technology, Government of India for a research project on Identification of adulterants in Indian spices.

Dr. Mandeep Singh is currently a Professor and Former Head in the Electrical and Instrumentation Engineering Department, Thapar Institute of Engineering & Technology, in Patiala, India. Dr. Singh obtained his Ph.D. in Tele-Cardiology, Master of Engineering in Computer Science, and Bachelor of Engineering in Electronics (Instrumentation and Control). Dr. Singh is a BEE Certified Energy Auditor and an Empaneled Consultant for PAT Scheme. Dr. Singh has more than 20 years of teaching experience. His current area of research interest includes Biomedical Instrumentation, Energy Conservation, Alternative Medicine, and Cognition Engineering. In addition to his regular responsibilities, he has served the Thapar Institute of Engineering & Technology as Faculty Advisor (Electrical) for more than eight years. Dr. Singh is currently handling two research projects with DIPAS-DRDO related to fatigue detection and wireless monitoring of ambulatory subjects.

Dr. Shruti Jain is an Associate Professor in the Department of Electronics and Communication Engineering at Jaypee University of Information Technology, in Waknaghat, H.P, India and has received her Doctor of Science (D. Sc.) degree in Electronics and Communication Engineering. She has 16 years of teaching experience and has filed five patents, out of which one patent is granted and four are published. She has published more than 15 book chapters, and 100 research papers in reputed indexed journals and in international conferences. She has also published six books. She has completed two government-sponsored projects. She has guided 6 Ph.D. students and now has 2 registered students. She has also guided 11 MTech scholars and more than 90 BTech undergrads. Her research interests are Image and Signal Processing, Soft Computing, Bioinspired Computing, and Computer-Aided Design of FPGA and VLSI circuits. She is a senior member of IEEE, life member and Editor-in-Chief of the Biomedical Engineering Society of India, and a member of the International Association of Engineers. She is a member of the editorial board of many reputed journals. She is also a reviewer of many journals and a member of the technical program committees of different conferences. She was awarded a Nation Builder Award in 2018–19.

Dr. Vishal Jain is presently working as an Associate Professor at the Department of Computer Science and Engineering, School of Engineering and Technology, Sharda University, in Greater Noida, U. P., India. Before that, he worked for several years as an Associate Professor at Bharati Vidyapeeth's Institute of Computer Applications and Management (BVICAM), in New Delhi. He has more than 14 years of experience in academics. He has earned several degrees: Ph.D. (CSE), MTech (CSE), MBA (HR), MCA, MCP, and CCNA. He has more than 370 research citations with Google Scholar (h-index score 9 and i-10 index 9) and has authored more than 70 research papers in professional journals and conferences. He has authored and edited more than 10 books with various publishers, including Springer, Apple Academic Press, CRC Press, Taylor & Francis Group, Scrivener, Wiley, Emerald, and IGI-Global. His research areas include information retrieval, semantic web, ontology engineering, data mining, ad hoc networks, and sensor networks. He received a Young Active Member Award for the year 2012–13 from the Computer Society of India, and Best Faculty Award for the year 2017 and Best Researcher Award for the year 2019 from BVICAM, New Delhi.

List of Contributors

A. Sheik Abdullah
Thiagarajar College of Engineering
Chennai, India

Hritam Basak
Jadavpur University
West Bengal, India

Arindam Chakrabarty
Department of Management,
    Rajiv Gandhi University (Central
    University)
Arunachal Pradesh, India

Soham Chattopadhyay
Jadavpur University
West Bengal, India

Uday Sankar Das
Department of Management &
    Humanities, National Institute
    of Technology
Arunachal Pradesh

Arijit Dey
Maulana Abul Kalam Azad University
    of Technology
West Bengal, India

A. Hridya
CHRIST Deemed to be University
Uttar Pradesh, India

Aditya Anand Doshi
MIT College of Engineering
Pune, India

Meenu Garg
Chitkara University Institute of
    Engineering and Technology,
    Chitkara University
Punjab, India

Ahona Ghosh
Maulana Abul Kalam Azad University
    of Technology
West Bengal, India

Deepali Gupta
Chitkara University Institute of
    Engineering and Technology,
    Chitkara University
Punjab, India

Sheifali Gupta
Chitkara University Institute of
    Engineering and Technology,
    Chitkara University
Punjab, India

Sowmya Kamath S
National Institute of Technology
    Karnataka
Surathkal, India

K. Karthik
National Institute of Technology
    Karnataka
Surathkal, India

P. Karthikeyan
Thiagarajar College of Engineering
Chennai, India

Utkarsh Nitin Kasara
MIT College of Engineering
Pune, India

Hafiz T.A. Khan
University of West London
London, England

Suchitra Khojeb
MIT College of Engineering
Pune, India

R. Priyatharshini
Easwari Engineering College
Tamil Nadu, India

Saket Kushwaha
Rajiv Gandhi University
  (Central University)
Arunachal Pradesh, India

Pavanalaxmi
Sahyadri College of Engineering &
  Management
Karnataka, India

Ashwani Kumar Mishra
All India Institute of Medical Sciences
  (AIIMS)
New Delhi, India

Rajiv Narang
All India Institute of Medical Sciences
  (AIIMS)
New Delhi, India

Dilip C. Nath
Assam University
Assam, India

Siddhanth Pillay
National Institute of Technology
  Karnataka
Surathkal, India

R. Parkavi
Thiagarajar College of
  Engineering
Chennai, India

Reena R. Roy
Easwari Engineering College
Tamil Nadu, India

Sriparna Saha
Maulana Abul Kalam Azad University
  of Technology
West Bengal, India

S. Selvakumar
GKM College of Engineering and
  Technology
Tamil Nadu, India

Tajinder Pal Singh
Chitkara College of Applied
  Engineering, Chitkara University
Punjab, India

Anita Verma
All India Institute of Medical Sciences
  (AIIMS)
New Delhi, India

Vivek Verma
All India Institute of Medical Sciences
  (AIIMS)
New Delhi, India

A. Vijayalakshmi
CHRIST Deemed to be University
Uttar Pradesh, India

Roopashree
Sahyadri College of Engineering &
  Management
Karnataka, India

1 Impact of Gender on the Lipid Profile of Patients with Coronary Artery Disease
A Bayesian Analytical Approach

Vivek Verma, Ashwani Kumar Mishra, Anita Verma, Hafiz T. A. Khan, Dilip C. Nath, and Rajiv Narang

CONTENTS

1.1   INTRODUCTION

Cardiovascular diseases are grouped into diseases based on problems of the heart and blood vessels [1], including acute coronary syndrome and coronary artery disease (CAD). CAD occurs when the heart has not received adequate amounts of oxygen and blood due to plaque buildup within the coronary arteries, and CAD is the most common type of heart disease. The deposition of cholesterol (known as atherosclerosis) and other materials (called plaque) within the coronary arteries causes the arteries to narrow and harden, which affects the blood supply to the heart. If such deposition continues, then it will lead to heart failure. According to the World Health Organization, CAD is one of the leading causes of morbidity and mortality and is also the leading cause of death.

1

According to the American Heart Association, among several risk factors for coronary heart disease, the components of the lipid profile, namely triglycerides, low-density lipoprotein cholesterol (LDL-C), direct high-density lipoprotein cholesterol (HDL-C), and total cholesterol are very common. According to the Centers for Disease Control and Prevention (CDC), CAD is the leading cause of death in the United States for both men and women. CAD alone is responsible for more than 4.5 million deaths worldwide [2].

Risk factors associated with CAD include lifestyle, environment, and genetic factors [3]. Previous CAD studies have documented the association of the lipid profile; in this sense, intensive lifestyle changes can also stop or reverse its progression without the use of lipid-lowering drugs [4]. According to the American Heart Association, the recommended normal range prescribed for a lipid profile is: total cholesterol < 200 mg/dL, triglycerides < 200 mg/dL, HDL-C > 40 mg/dL, and LDL-C < 130 mg/dL. The lipid profile acts as a diagnostic tool for the detection and clinical management of cases of CAD. For example, in dyslipidemia, patients with CAD, LDL-C, total cholesterol and triglycerides, will be higher, and HDL-C cholesterol will be lower [5]. It is also well known that in patients with CAD, triglycerides [6], total cholesterol, and LDL-C are significantly higher, and HDL-C is significantly lower [7–10].

The influence of gender on the components of the lipid profile is also studied in various situations. Total cholesterol levels in women are significantly reduced in the 25- to 49-year age group and are higher in the 50- to 64-year age group than in men [11]. Total cholesterol, LDL-C, and the ratio of total cholesterol to HDL-C levels are significantly higher in older women (> 50 years) than in younger women (30–46 years), but in men, these levels do not change dramatically with age [12]. The impact of gender on triglycerides turns out to be significantly different [13–15].

In a study [6], it had shown that 13.3% of the population older than 55 years were affected by CAD and, among them, the percentage of men was higher than that of women. The appearance of CAD is associated with changes in the lipid profile, which is influenced by several factors, and among them, the sex of an individual is also an important factor. Therefore, the objective of the present study is to examine the impact of sex on the lipid profile in patients with CAD. For this study, the data set from the National Health and Nutrition Examination Survey (NHANES), 2015–2016, was used. The study population is made up of people 50 years of age or older. To compare the differences in the lipid profile between the sexes when the sample size is comparatively small, both a classic two-tailed Student test and a non-parametric Wilcoxon rank sum test, as well as the Bayesian t-test were adopted. Statistical significance was measured using p-values in the context of the Student's t-test and the Wilcoxon rank sum test, while the Bayes factor was used for the Bayesian t-test.

1.2   METHODS

For the evaluation of the health and nutrition component of the noninstitutionalized population in the United States, since the early 1960s, the CDC has conducted the NHANES. The NHANES program was initiated to assess the level of health and nutritional status in the United States, collecting information on various characteristics—household, physical, and medical examinations of sampled children and adults. In this study, we used NHANES 2015–2016, which was launched with 15,327 people. The NHANES (2015–2016) database was considered, which included 9,971 individuals, who completed the interview.

1.2.1   STUDY POPULATION

Participants aged less than 50 ($n = 8635$) and lipid profiles were not observed or declared ($n = 291$), are excluded. There were 1,045 participants, aged 50 years or older, and among them, 91 were clinically diagnosed with CAD and the remaining 945 did not have CAD (non-CAD), and their lipid profiles were observed and reported.

1.2.2   LABORATORY METHODS

Information on laboratory parameters, including lipid profiles, including triglycerides (mg/dL), LDL cholesterol (mg/dL), direct HDL cholesterol (mg/dL), and total cholesterol (mg/dL), were obtained from the participants, who were recommended too fast for at least nine hours before physical examination at the mobile examination center (MEC) for blood collection.

The criteria of the lipid standardization program of the CDC were used to standardize the parameters of the serum lipid profile due to changes in laboratory methods during years of research to ensure accuracy and comparability of measurements between studies.

1.3   STATISTICAL ANALYSIS

For the comparison of the descriptive statistics among gender, the results were expressed as Mean (μ) $\pm$ Standard Deviation (s) and percentage (%). Under the assumption that the variation among the components of lipid profile and other continuous differences based on sex (for male (μ_M, s_M); female (μ_F, s_F)) are fixed quantities, to test the hypothesis

$$H_0 : \mu_M = \mu_F \text{ vs } H_1 : \mu_M \neq \mu_F \tag{1.1}$$

the classical two-tail Student's t-test and Wilcoxon rank sum test under parametric and nonparametric setup, respectively, were appropriately used and discussed, and statistical significances were measured using their p-values. For testing the hypothesis of equation (1.1), the test statistic takes the following form under the classical paradigm:

$$t = \frac{\overline{x}_M - \overline{x}_F}{\left(\dfrac{(n_M - 1)s_M^2 + (n_F - 2)s_F^2}{n_M + n_F - 2} \right)^{1/2} \Big/ \sqrt{n_\theta}} \tag{1.2}$$

where, $n_\theta = \left(\frac{1}{n_M} + \frac{1}{n_F}\right)^{-1}$, the degrees of freedom are $\tau = n_M + n_F - 2$, $\bar{x}_i$ and μ_i respectively, denotes the sample and population mean corresponding to each the continuous quantity of the ith gender, $\{i = Male\,(M),\, Female\,(F)\}$.

In real sense, the exact characterization of the randomness inherent in the quantitative measurement is ignored. Under such situation, the comparison of any continuous quantities and their assessments under a traditional test of significance becomes a serious concern. Therefore, the present work emphasizes another promising paradigm of the statistical framework that can address such a situation by considering the formulation under the Bayesian t-test. In this analytical procedure, a reasonable and useful prior has suggested to obtain a closed form of Bayes factor for emphasizing the statistical significance. To test the hypothesis under two-sided alternatives, the Bayesian version of the two-sample t statistic under the null and alternative hypotheses was adopted, and the decision was made using the value of Bayes factor (B). In the present study, common variance, say σ^2, has assumed both sex corresponding to each quantity. In order to work with the Bayesian paradigm, we need to specify the prior distribution of the effect size (difference) that needs to be tested. Under the suggested hypothesis of a nonzero difference, the standardized difference $\frac{|\mu_M - \mu_F|}{\sigma}$ has prior mean, say θ, and prior variance, say σ_θ^2. The Bayes factor for testing H_0 against H_1 of equation (1.1) is:

$$B(x) = \frac{T_\tau(t \mid 0,1)}{T_\tau(t \mid \theta\sqrt{n_\theta}, 1 + n_\theta\,\sigma_\theta^2)} \tag{1.3}$$

where $T_\tau(t \mid \alpha, \beta)$ denotes the value that results from plugging t into noncentral t distribution probability density function with f degree of freedom and parameters α for location and $\beta^{1/2}$ for scale [16]. The rule of thumb [17–18] followed for inference is as follows, if $log_{10}(B(x))$ varies between 0 and 0.5, the evidence against null hypothesis H_0 will be poor, if $log_{10}(B(x))$ lies between 0.5 and 1, it is substantial, if it is between 1 and 2, it is strong, and if it is above 2 it is decisive. The results are simulated by following the Gibbs sampling, with 100,000 iterations, by using R-software version 3.6.2, and data processing is done using the SAS University edition.

1.4   RESULTS

1.4.1   DESCRIPTIVE CHARACTERISTICS

Table 1.1 shows the comparison of patients with CAD and non-CAD, which includes the age at which CAD occurred, its duration, if the doctor ever said the person was obese and/or to reduce salt and fat/calories intake. A total of 91 CAD and 954 non-CAD participants [males (CAD = 55; non-CAD = 450) and females (CAD = 36; non-CAD = 504)] were included in this study. The overall mean age ± general standard deviation of the participants was 69.8 ± 7.5 years in CAD (54–80 years) and 64.5 ± 9.2 in non-CAD (50–80) years. The mean age of presentation to seek

TABLE 1.1

Demographic and Clinical Characteristics of the Patients with CAD Versus No CAD

Characteristics and Categories		CAD (*n* = 91) (Mean ± SD/ Percentage) (Range)	No CAD (*n* = 954) (Mean ± SD/ Percentage) (Range)
Age (in years)	Male	69.3 ± 7.2 (54–80)	64.5 ± 8.9 (50–80)
(Mean ±SD) (range)	Female	70.6 ± 7.9 (57–80)	64.5 ± 9.4 (50–80)
	All	69.8 ± 7.5 (54–80)	64.5 ± 9.2 (50–80)
CAD Occurrence Age (in years)	Male	57.2 ± 9.8	NA
(Mean ±SD)	Female	60.4 ±10.5	NA
	All	58.5 ± 10.2	NA
	Male	12.2 ± 8.1	NA
Duration of period of CAD	Female	10.1 ± 8.9	NA
	All	11.4 ± 8.4	NA
"Doctor ever said you were overweight"*	Yes	47 (51.7)	370 (38.8)
"Doctor told to reduce salt in diet" *	Yes	48 (52.8)	350 (36.7)
"Doctor told to reduce fat/calories"*	Yes	47 (57.7)	371 (38.9)

*From the questionnaire used in the study.

treatment for CAD in men (57.2 years) was earlier than in women (60.4 years). The duration of CAD in men (12.2 years) was longer than in women (10.1 years). Most of the patients with CAD were obese (51.7%), and it was recommended to reduce their salt intake (52.8%) and diet control (57.8%). Among non-CAD participants, the percentage distribution of obese people prescribed to reduce salt intake and diet control is almost the same.

1.4.2 CLINICAL FEATURES

A comprehensive gender comparison between the lipid profile parameters as well as some of the derived parameters, namely triglycerides (mg/dL), LDL cholesterol (mg/dL), direct HDL cholesterol (mg/dL) and total cholesterol (mg/dL), non-HDL cholesterol, TC: HDL and LDL: HDL ratio are listed in Tables 1.2–1.4. The gender association between the various components of the lipid profile and their derived proportions has been measured under the classical (both parametric and nonparametric) and Bayesian paradigms. Furthermore, the empirical gender-wise distribution pattern of each of the lipid parameters and their means are shown in Figures 1.1–1.2, participants without CAD (*n* = 945) and CAD participants (*n* = 91).

In Table 1.2, it was hypothesized (null hypothesis) that there is no gender difference in the lipid profile of the participants (independent of CAD and absence of CAD). The classical t-test and nonparametric Wilcoxon rank sum test based

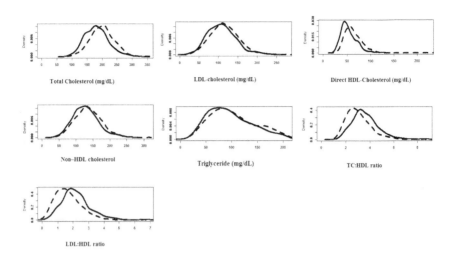

FIGURE 1.1 Gender-wise empirical distributional patterns of the lipid parameters pattern of no CAD participants (Male = 450; Female = 504), black denotes male and dashed denotes female.

on participants (n = 1,045) suggested a significant difference between the sexes ($p < 0.05$) for triglycerides, LDL cholesterol, direct HDL cholesterol, total cholesterol, and non-cholesterol, HDL, TC:HDL ratio and LDL:HDL ratio of the lipid profile. The similar significant gender difference in lipid profiles was also captured by Bayesian t-tests and revealed that the Bayes factor (Log (B)) is greater than 2

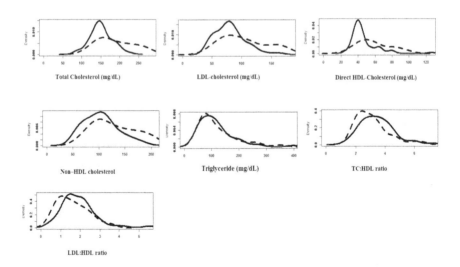

FIGURE 1.2 Gender-wise empirical distributional patterns of the lipid parameters pattern of CAD participants (Male = 55; Female = 36), black denotes male and dashed denotes female.

TABLE 1.2
Classical and Bayesian Evaluation of Gender Differences in Association of Lipid Profile Among Total Patients (Large Sample Size)

Parameter	Gender (n = 1,045)		Parametric Test for Difference in Mean		Nonparametric Wilcoxon rank sum test	Bayes Factor (Log(B))
	Male (n_M = 505) Mean ± SD	Female (n_F = 540) Mean ± SD	t-value	p-Value	p-Value	
Total Cholesterol (mg/dL)	178.80 ± 39.91	200.30 ± 40.97	−8.56	< .0001	< .0001	14.03
LDL Cholesterol (mg/dL)	105.90 ± 36.25	115.80 ± 36.54	−4.39	< .0001	< .0001	2.76
Direct HDL Cholesterol (mg/dL)	53.02 ± 16.73	63.41 ± 18.93	−9.37	< .0001	< .0001	16.96
Non-HDL Cholesterol	125.80 ± 38.28	136.80 ± 39.68	−4.57	< .0001	< .0001	5.11
Triglyceride (mg/dL)	99.21 ± 46.07	105.00 ± 46.66	−2.00	0.0457	0.0347	8.58
TC:HDL ratio	3.60 ± 1.11	3.36 ± 1.03	3.69	0.0002	< .0001	2.73
LDL:HDL ratio	2.17 ± 0.94	1.98 ± 0.85	3.40	0.0007	0.0005	2.29

as triglycerides (Log (B) = 8.58), LDL cholesterol (Log (B) = 2.76), direct HDL cholesterol (Log (B) = 16.96), total cholesterol (Log (B) = 14.03) and non-HDL cholesterol (Log (B) = 5.11), TC: HDL ratio (Log (B) = 2.73) and LDL: HDL ratio (Log (B) = 2.29).

The classical t-test and nonparametric Wilcoxon rank sum test based on no-CAD participants (n = 945) is presented in Table 1.3, which was also suggesting a significant (p < 0.05) gender-wise difference among Triglyceride, LDL cholesterol, direct HDL cholesterol, total cholesterol and non-HDL cholesterol, TC:HDL ratio and LDL:HDL ratio of lipid profile. Significant differences among lipid profiles of males and females were also captured by Bayesian t-tests with Bayes factor (Log(B)) greater 2, that is, triglyceride (Log(B) = 4.76), LDL cholesterol (Log(B) = 2.45), direct HDL cholesterol (Log(B) = 32.62), total cholesterol (Log(B) = 22.79) and non-HDL cholesterol (Log(B) = 3.51), TC:HDL ratio (Log(B) = 4.54) and LDL:HDL ratio (Log(B) = 4.06).

The classical t-test and nonparametric Wilcoxon rank sum test based on CAD participants (n = 91) is presented in Table 1.4, which suggested significant gender-wise differences only among LDL cholesterol, direct HDL cholesterol, total cholesterol, and non-HDL cholesterol, of lipid profile. On the other hand, Bayesian

TABLE 1.3

Classical and Bayesian Evaluation of Gender Differences in Association of Lipid Profile Among Patients with No CAD (Large Sample Size)

Parameter	Gender ($n = 954$)		Parametric Test for Difference in Mean		NonParametric Wilcoxon Rank Sum Test	Bayes Factor (Log(B))
	Male ($n_M = 450$) Mean ± SD	Female ($n_F = 504$) Mean ± SD	t-Value	p-Value	p-Value	
Total Cholesterol (mg/dL)	182.20 ± 39.67	201.10 ± 40.29	−7.28	< .0001	< 0.0001	22.79
LDL Cholesterol (mg/dL)	109.00 ± 36.14	116.60 ± 36.26	−3.23	0.0013	0.0006	2.45
Direct HDL Cholesterol (mg/dL)	53.76 ±16.62	63.72 ± 18.84	−8.61	< .0001	< 0.0001	32.62
Non-HDL Cholesterol	128.50 ± 38.04	137.40 ± 39.51	−3.55	0.0004	0.0003	3.51
Triglyceride (mg/dL)	97.12 ± 42.47	103.80 ± 44.28	−2.38	0.0176	0.0193	4.76
TC:HDL ratio	3.61 ± 1.10	3.35 ± 1.02	3.81	0.0001	< 0.0001	4.54
LDL:HDL ratio	2.20 ± 0.94	2.00 ± 0.86	3.68	0.0003	0.0001	4.06

t-tests suggested significant gender-wise differences among all lipid profile parameters with Bayes factor (Log(B)) greater 2, that is, triglyceride (Log(B) = 7.21), LDL cholesterol (Log(B) = 2.52), direct HDL cholesterol (Log(B) = 3.36), total cholesterol (Log(B) = 3.38) and non-HDL cholesterol (Log(B) = 4.55), TC: HDL ratio (Log(B) = 4.55) and LDL: HDL ratio (Log(B) = 3.76).

1.5   DISCUSSION

Table 1.1 shows that the mean age of participants with CAD is higher than that of non-CAD participants. The mean age of onset of CAD in men was lower than that of women; therefore, the mean duration of the CAD period was longer in men than in women. The age factor is an important predictor for CAD. The majority of the CAD prevalence occurred between the ages of 50 and 70, approximately four times the prevalence in people older than 70 years. Among patients with CAD compared to non-CAD, more than 50% of the individuals corresponding to each of the risk factors of being overweight, high salt intake, and high fat/calorie intake were prescribed to reduce the intake and food control. This suggests the need to focus on the daily routine of the participants.

TABLE 1.4

Classical and Bayesian Evaluation of Gender Differences in Association of Lipid Profile Among Patients with CAD (Small Sample Size)

Parameter	Gender ($n = 91$)		Parametric Test for Difference in Mean		Nonparametric Wilcoxon Rank Sum Test	Bayes Factor (Log(B))
	Male ($n_M = 55$) Mean ± SD	Female ($n_F = 36$) Mean ± SD	t-Value	p-Value	p-Value	
Total Cholesterol (mg/dL)	151.02 ±29.92	188.58 ±48.37	−4.15	< .0001	0.0004	3.38
LDL Cholesterol (mg/dL)	80.80 ±26.14	105.28 ±39.24	−3.29	0.0017	0.0039	2.52
Direct HDL Cholesterol (mg/dL)	46.96 ±16.60	59.14 ±19.88	−3.16	0.0021	0.0007	3.36
Non-HDL cholesterol	104.10 ±95.08	129.40 ±115.30	−3.21	0.0018	0.0054	4.55
Triglyceride (mg/dL)	116.29 ±66.93	120.8 ±71.31	−0.30	0.7613	0.8806	7.21
TC:HDL ratio	3.50 ±1.17	3.43 ±1.15	0.24	0.8110	0.5976	6.14
LDL:HDL ratio	1.91 ±0.90	1.91 ±0.79	−0.01	0.9893	0.9418	3.76

Nonsignificant results obtained in both of the classic nonparametric Wilcoxon rank sum t-tests for some of the important lipid parameters, namely triglycerides, TC:HDL ratio and LDL:HDL ratio, which is considered a good predictor of CAD, was found significantly different in Tables 1.2 and 1.3, which suggests that the null hypothesis is contrary to the theory of the difference between the sexes, and is also observed in the empirical densities shown in Figure 1.1–1.2.

Previous studies that focused on the impact of triglycerides, the TC:HDL ratio, and the LDL:HDL ratio on CAD have shown that elevated triglyceride levels increase the risk of prevalence of coronary heart disease and is lowered through clinical management in addition to diet control, regular exercise, and pharmacotherapy [19]. This is of great importance for public health since such a suggestion can have positive reinforcement among patients toward adopting a healthy dietary pattern in their daily routine. The higher value of the LDL:HDL ratio shows a positive association with the prevalence of hypertension and hypercholesterolemia in men and women [20] and the higher TC:HDL ratio was considered an independent indicator of extensive coronary disease [21]. As with the classic t-test and nonparametric Wilcoxon rank sum paradigms, some of the important lipid parameters, namely triglycerides, the TC:HDL ratio, and the LDL:HDL ratio were not found to differ significantly across gender, which were found to be different in earlier studies. Based on the results obtained, data-based estimates for lipid profile parameters were found to be consistent with clinical characteristics and were also found to be effective in demonstrating statistical significance with clinical significance. Therefore, the quality of the data was not questioned regarding the insensitivity to distinguish the theory from the null hypothesis. However, to clinically link the data to theory, apart from certain lipid parameters, namely LDL cholesterol, direct HDL cholesterol, total cholesterol, and non-HDL cholesterol, these were found to be nonsignificant in the classical tests, Bayesian technique was adopted in Table 1.4. The Bayesian t-test suggested evidence of differences in the lipid profile across gender and was also observed in the empirical densities shown in Figure 1.1. The significant difference in elevated levels of lipid parameters, namely triglycerides, LDL cholesterol, direct HDL cholesterol, and total cholesterol in women with CAD as compared to men, has also been discussed in several other studies [22–26], which has also been observed under the Bayesian test paradigm.

The lipid parameters of LDL cholesterol, direct total HDL cholesterol, and non-HDL cholesterol that were significantly different between sexes according to the classical test paradigms also corresponded to the Bayesian paradigms. On the other hand, the reverse is not true, as triglycerides, TC:HDL ratio, and LDL:HDL ratio also differed significantly across gender in Bayesian t-tests and were discussed in previous studies, but they were not captured in the conventional classical tests, possibly due to a small sample size.

1.6 CONCLUSION

In the study, the Bayesian inferential procedure is presented, where the sample size is comparatively smaller, with emphasis on the possible differences in the parameters of the lipid profile of patients with CAD between men and women. Assuming that the differences in parameters due to gender are fixed, the classical t-test and the nonparametric Wilcoxon rank sum test were not fully compatible to capture significant changes in lipid parameters due to gender. On the other hand, even with a small sample size, the results obtained on the basis of Bayesian t-tests turned out to be more reliable for concordance of clinical practices on the sex difference in the association of lipid profile in patients. Patients with CAD whose results were not fully recognized in the conventional t-tests and Wilcoxon rank sum nonparametric tests, viz. Triglycerides (p-value = 0.7613 (0.8806), Log (B) = 7.21), TC:HDL ratio (p-value = 0.8110 (0.5976), Log (B) = 6, 14) and LDL:HDL ratio (p-value = 0.9893 (0.9418), Log (B) = 3.76).

REFERENCES

1. Sanchis-Gomar, F., Perez-Quilis, C., Leischik, R., & Lucia, A. Epidemiology of coronary heart disease and acute coronary syndrome. *Annals of Translational Medicine*, *4*(13), 2016.
2. Okrainec, K., Banerjee, D. K., & Eisenberg, M. J. Coronary artery disease in the developing world. *American Heart Journal*, *148*(1), 7–15, 2004

3. Musa, H. H., Tyrab, E. M., Hamid, M. M., Elbashir, E. A., Yahia, L. M., & Salih, N. M. Characterization of lipid profile in coronary heart disease patients in Sudan. *Indian Heart Journal*, 65(2), 232–233, 2013. doi:10.1016/j.ihj.2013.03.007

4. Ornish, D., Scherwitz, L. W., Billings, J. H., Gould, K. L., Merritt, T. A., Sparler, S., ... & Brand, R. J. Intensive lifestyle changes for reversal of coronary heart disease. *JAMA*, 280(23), 2001–2007, 1998

5. Haque, A. E., Yusoff, F. B. M., Ariffin, M. H. S. B., Fadhli, M., Ab Hamid, B., Hashim, S. R. B., & Haque, M. Lipid profile of the coronary heart disease (CHD) patients admitted in a hospital in Malaysia. *Journal of Applied Pharmaceutical Science*, 6(05), 137–142, 2016.

6. Kumar, L., & Das, A. L. Assessment of serum lipid profile in patients of coronary artery disease: a case-control study. *International Journal of Contemporary Medical Research*, 5(5), 59–62, 2018.

7. Haddad, F. H., Omari, A. A., Shamailah, Q. M., Malkawi, O. M., Shehab, A. I., Mudabber, H. K., & Shubaki, M. K. Lipid profile in patients with coronary artery disease. *Saudi Medical Journal*, 23(9), 1054–1058, 2002.

8. Bestehorn, K., Jannowitz, C., Karmann, B., Pittrow, D., & Kirch, W. Characteristics, management and attainment of lipid target levels in diabetic and cardiac patients enrolled in Disease Management Program versus those in routine care: LUTZ registry. *BMC Public Health*, 9(1), 280, 2009.

9. Abraham, G. Evaluation of variation in the lipid profile and risk for coronary artery disease in healthy male individuals with respect to age. *International Journal of Research Medicine Science*, 2(2), 551–556, 2014.

10. Sun, X., & Du, T. Trends in cardiovascular risk factors among US men and women with and without diabetes, 1988–2014. *BMC Public Health*, 17(1), 893, 2017.

11. Gostynski, M., Gutzwiller, F., Kuulasmaa, K., Döring, A., Ferrario, M., Grafnetter, D., & Pajak, A. Analysis of the relationship between total cholesterol, age, body mass index among males and females in the WHO MONICA Project. *International Journal of Obesity*, 28(8), 1082, 2004.

12. Goh, V. H., Tong, T. Y., Mok, H. P., & Said, B. Differential impact of aging and gender on lipid and lipoprotein profiles in a cohort of healthy Chinese Singaporeans. *Asian Journal of Andrology*, 9(6), 787–794, 2007.

13. Rosenkranz, S. K. Lifestyle influences on airway health in children and young adults (Doctoral dissertation, Kansas State University), 2010.

14. Carroll, M. D., Kit, B. K., & Lacher, D. A. *Trends in elevated triglyceride in adults: United States, 2001–2012 (No. 2015)*. US Department of Health and Human Services, Centers for Disease Control and Prevention, National Center for Health Statistics, 2015.

15. Alzahrani, S. H., Baig, M., Aashi, M. M., Al-shaibi, F. K., Alqarni, D. A., & Bakhamees, W. H. Association between glycated hemoglobin (HbA1c) and the lipid profile in patients with type 2 diabetes mellitus at a tertiary care hospital: a retrospective study. *Diabetes, metabolic syndrome and obesity: targets and therapy*, 12, 1639, 2019.

16. Gronau, Q. F., Ly, A., & Wagenmakers, E. J. Informed Bayesian t-tests. *The American Statistician*, 1–14, 2019.

17. Kass, R. E., & Raftery, A. E. Bayes Factors. *Journal of the American Statistical Association*, 90(430), 773, 1995.

18. Verma, V., Mishra, A. K., & Narang R. Application of Bayesian analysis in medical diagnosis. *Journal of Practice of Cardiovascular Science*, 5,136–41, 2019.

19. Coughlan, B. J., & Sorrentino, M. J. Does hypertriglyceridemia increase risk for CAD? Growing evidence suggests it plays a role. *Postgraduate Medicine*, 108(7), 77, 2000.

20. Chen, Q. J., Lai, H. M., Chen, B. D., Li, X. M., Zhai, H., He, C. H., ... & Ma, Y. T. Appropriate LDL-C-to-HDL-C ratio cutoffs for categorization of cardiovascular

disease risk factors among Uygur adults in Xinjiang, China. *International Journal of Environmental Research and Public Health*, *13*(2), 235.0, 2016.

21. Luz, P. L. D., Favarato, D., Faria-Neto Junior, J. R., Lemos, P., & Chagas, A. C. P. High ratio of triglycerides to HDL-cholesterol predicts extensive coronary disease. *Clinics*, *63*(4), 427–432, 2008.

22. Schildkraut, J. M., Myers, R. H., Cupples, L. A., Kiely, D. K., & Kannel, W. B. Coronary risk associated with age and sex of parental heart disease in the Framingham Study. *The American Journal of Cardiology*, *64*(10), 555–559, 1989.

23. Brown, S. A., Morrisett, J. D., Boerwinkle, E., Hutchinson, R., & Patsch, W. The relation of lipoprotein [a] concentrations and apolipoprotein [a] phenotypes with asymptomatic atherosclerosis in subjects of the Atherosclerosis Risk in Communities (ARIC) Study. *Arteriosclerosis and Thrombosis: A Journal of Vascular Biology*, *13*(11), 1558–1566, 1993.

24. Bostom, A. G., Gagnon, D. R., Cupples, L. A., Wilson, P. W., Jenner, J. L., Ordovas, J. M., … & Castelli, W. P. A prospective investigation of elevated lipoprotein (a) detected by electrophoresis and cardiovascular disease in women. The Framingham Heart Study. *Circulation*, *90*(4), 1688–1695, 1994.

25. Roeters van Lennep, J. E., Westerveld, H. T., Erkelens, D. W., & van der Wall, E. E. Risk factors for coronary heart disease: implications of gender. *Cardiovascular Research*, *53*(3), 538–549, 2002.

26. Hossain, A., & Khan, H. T. Risk factors of coronary heart disease. *Indian Heart Journal*, *59*(2), 147–151, 2007.

2 Implementation of Wearable ECG Monitor Interfaced with Real-Time Location Tracker

Aditya Anand Doshi, Utkarsh Nitin Kasar, and Suchitra Khoje

CONTENTS

2.1   INTRODUCTION

Breakthroughs have been made in the last decade in the field of wearable wireless health-monitoring systems (WWHMS), and the wearable technology industry is growing at a staggering compound annual growth rate (CAGR) of 19% and will reach $54 billion by 2023 [1–4]. Because of the increasing population of the elderly, that is, people older than 60 (841 million in 2013 worldwide), and due to tremendous leaps in medical science, this population may surpass 2 billion by 2050. Nevertheless, due to the slow but steady rise in healthcare costs, nations are now facing cost complications in providing long-term health care to their citizens [5–8]. Yet 31% of all global deaths occur due to cardiovascular diseases, 85% of which are caused by heart attack and stroke [9]. The impetus for the evolution and manufacturing in the domain of wearable sensors and devices for the healthcare sector is, therefore, immense [10–12], and the advantages could be correlated with the long-term monitoring of a hospital patient in their personal environment. To meet this requirement, researchers believe that long-term monitoring [13–17] of physiological data may lead to significant diagnostic and treatment improvements. This has resulted in the development of several device prototypes and commercial products [18–20] and miniaturization of wearable devices in recent years [21], the goal of which is to provide real-time feedback on a patient's health data, to the patient or a medical center or directly to a qualified supervisor, while being able to alert the person if conditions threatening their health are imminent [22]. This chapter proposes a wearable ECG monitoring system integrated with real-time location tracking, which meets the requirement of continuous health monitoring of the patient by transmitting a distress signal containing the location and ECG signal data of the patient having the medical emergency to the emergency medical services (EMS) team of the hospital.

The popular use of mobile communication networks, and its wide coverage, along with the popularization of software and Internet of Things (IoT)–based information systems in hospitals has been taken advantage of by our system. Patients can easily make use of the mobile monitoring equipment of our system to acquire physiological data and upload this collected data to the central database, provided that the system is connected to the internet.

2.2   HARDWARE ARCHITECTURE OF WEARABLE USER-END DEVICE

2.2.1   ARCHITECTURE OF SYSTEM HARDWARE

Our device is designed to be a wearable medical device that should be on the person, at all times. Therefore, the system architecture is designed to be as lightweight and less bulky as possible [23]. The implementation of this lifesaving device is achieved with a minimum number of hardware electronic components.

The system is divided into four major subsystems as shown in Figure 2.2—mainly, a wearable device that is placed on the patient's body to capture their real-time location (an ECG signal), a patient's mobile phone app that receives the lifesaving information from the wearable hardware setup and uploads it to an online patient data storage platform, an EMT/doctor's mobile phone app that downloads patient information and displays the location and ECG report of the patient facing the medical emergency, and a database. The central hardware composition consists of a power supply module, a microprocessor, an ECG acquisition module, a location-tracking GPS module, a Bluetooth (HC-05) module, etc.

The wearable device is equipped with a power supply circuit that drives power from a rechargeable lithium-ion polymer (LiPo) battery. The power supply circuit is designed to provide a regulated DC power supply to all the electronic components and special purpose modules within their recommended voltage range. It includes voltage divider and voltage regulator circuitry to ensure the safety of the patient and proper operation of the onboard electronic components. The use of rechargeable LiPo batteries makes the wearable device more efficient and portable. Also, the LiPo batteries are lightweight and have a lower chance of suffering from leaking electrolyte, thus, making it safe for use in our wearable device.

We have used the Arduino Uno [24] microcontroller board in our wearable device, which uses the ATmega328 Integrated Circuit (IC). The microcontroller board has 14 digital input/output (I/O) pins. It consists of a USB connection, making it very convenient to connect to a computer or a laptop, making connections easy. It has a recommended input voltage of 7–12 V and provides two supply voltages of 5 V and 3.3 V on the board, which is suitable for both the sensors in our wearable device. ATmega328 has sufficient computing capacity to process the data generated from various modules that are onboard the wearable device, simultaneously. Arduino is cost-effective and has a simple programming Arduino IDE, which makes it an appropriate choice for our device.

The HC-05 module [25] follows Bluetooth Serial Port Profile (SPP) and is designed for a transparent and effective wireless serial data transfer. The HC-05 module has an operating voltage range of 3.3V to 5V and an operating current of 30 mA. It has up to 10–100 meters of range and can operate in both master as well as slave mode. USART is an interface with a programmable baud rate and supports the following baud rates: 9,600, 19,200, 38,400, 57,600, 115,200, 230,400, 460,800. The module has an integrated antenna and an edge connector. The module auto-connects to the last device on power by default, and in case of disconnection due to discrepancy in range it tries to automatically reconnect when back in range

The AD8232 [26,27] is primarily an integrated signal conditioning block for an electrocardiogram (ECG) among biopotential-medical measurement applications. The design of AD8232 has been construed in such a way to extract, amplify, and even filter out small biopotential signals in the presence of noisy conditions, which can generally be created by body movement or remote electrode placement. This sensor is a compact and a frugal, yet accurate, way to measure the electrical activity of the heart. This electrical activity can be charted as an ECG plot. ECG signals naturally contain a lot of noise elements within; here the AD8232 acts as an operational amplifier to filter and extract a noise-free signal with distinct P, QRS, and T waves. It is a fully integrated ECG front end with two- or three-electrode configurations and a single-supply operation voltage ranging between 2.0 V and 3.5 V. Along with excellent filtering, AD8232 provides lead-off detection with AC or DC options. It directly interfaces with data acquisition and analog-to-digital converters (ADCs) making it easier to use and incorporate in wearable medical devices.

The GPS module is based on the NEO-6M chip from u-blox [28]. This unit provides an excellent and accurate GPS location. It consists of a larger inbuilt 25 mm by 25 mm active GPS antenna, and the module supports the baud rate from 4,800 bits per second (bps) to 230,400 bps. A battery is included in the module for power backup and to extract a GPS lock faster. It can track up to 22 satellites on 50 channels to identify user locations anywhere in the world and comparatively has the highest level of sensitivity (i.e., −161 dB) tracking and consumes only 45 mA of supply current. It provides up to five location updates per second with a horizontal position accuracy of 2.5 meters. One of the major advantages of this chip is that it comes with a power save mode (PSM), which helps decrease the power consumption of the system; this is done by selectively switching parts of the receiver on and off. The power consumption of the module is hence reduced to a very low value of 11 mA, making it suitable for use in our wearable medical device.

The SIM800L GSM/GPRS module is a miniaturized global system for mobile communications (GSM) modem that achieves most of the functionality of a normal cell phone, such as sending SMS text messages, making or receiving phone calls, sending and receiving GPRS data (TCP/IP, HTTP), etc. The module supports Quad-band—GSM850, EGSM900, DCS1800, and PCS1900 network—hence, it works almost everywhere in the world. This module functions on a SIM800L GSM cellular chip from SIMCom. The chip has an operating voltage value from 3.4 V to 4.4 V, and hence it is ideal for operating the chip on a rechargeable, compact battery in a wearable device setup. A baud rate ranging from 1,200 bps to 115,200 bps is supported by the module. It uses a serial-based attention (AT) command set and accepts a micro SIM card. To establish a network connection, the module makes use of a compact external antenna. The board also has a UFL connector facility to keep the antenna away from the board, which provides additional flexibility for our system design. This makes the SIM800L GSM/GPRS module a perfect and compact solution for our patient location transmission needs via SMS text messages in case of a lack of internet connection around the patient. Thus, it plays a vital role in the "fail-safe" measure of our device by sending real-time location updates after every two minutes to the EMT/doctor in case there is an issue with the internet connectivity in the patient's mobile phone or any other practical problem.

2.2.2 IMPLEMENTATION AND WORKING OF ECG ACQUISITION AND A REAL-TIME LOCATION MODULE

The implementation of our device is carried out in two simultaneous operations, and the flow of the operation is given in Figure 2.1. The wearable device is in the power on mode when it is placed on the patient's body. When the patient suffers from an emergency of any nature, they press the SOS button on the wearable device. As soon as the SOS button is pressed, the wearable device comes out of the standby mode and performs two operations simultaneously.

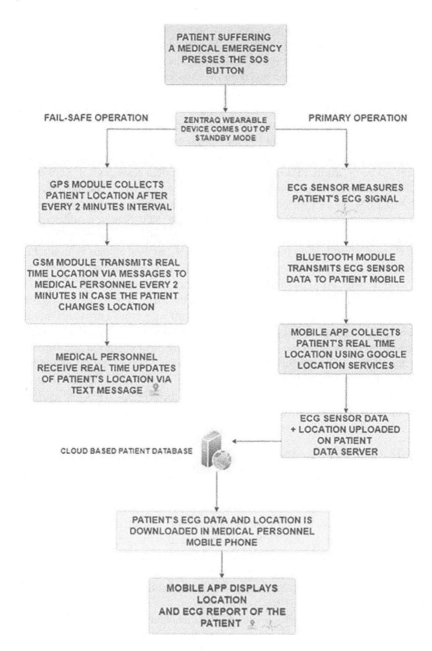

FIGURE 2.1   System Overflow.

2.2.2.1   Primary Operation

In this, the ECG sensor on the wearable device starts measuring the patient's ECG signal. The analog data of the ECG signal is sent to the microcontroller through the analog pin connection at A0. The onboard analog to digital converter of the microcontroller converts the analog ECG signal into a digital signal and sends it to the Bluetooth module. Thereafter, the Bluetooth module in the wearable device transmits the patient's ECG signal data to the mobile phone using serial data communication.

Once the patient's data is received on the mobile phone, the Zentraq patient app fetches the location of the patient using the mobile phone's GPS service. Then the app uploads the patient's location and ECG signal data on the Cloud Patient Database using the internet connection from the patient's mobile phone. The GPS location and a good internet connection from the mobile phone are very important for the successful completion of the primary operation and must be switched on at all times to avoid any type of failure in the operation. After the data is successfully uploaded on the Cloud Patient Database, the patient's location and the ECG data are available on the Zentraq EMT/doctor's mobile phone app. The patient's location can be accessed instantaneously by simply clicking on the GET LOCATION button in the app.

By pressing the GET LOCATION button, the patient's location is directly opened in the Google Maps app on the medical personnel's mobile phone. With the help of this app, they can navigate to the patient's location quickly and easily without losing precious time, which will be lifesaving for the patient in an emergency. Along with the patient's location, the Zentraq app also provides the medical personnel using the app with the ECG report/signal of the patient who is suffering from a medical emergency.

This ECG report is a vital piece of information that will be very helpful in constructing a preliminary diagnosis of the patient's medical condition and emergency. Also, the ECG report will help the medical personnel to prepare all the necessary arrangements and procure special equipment, etc., that will be needed to treat the patient properly, even before the patient arrives at the hospital. Thus, the availability of crucial information of the patient to the medical personnel will, in turn, save precious time in the golden hours, and the patient's life.

2.2.2.2   Fail-Safe Operation

The wearable device consists of a fail-safe model that sends real-time location updates of the patient facing a medical emergency, in case the primary operation model of the Zentraq device fails to function properly due to certain issues. The wearable device has an onboard GPS module that is activated as soon as the patient presses the SOS button. This GPS module calculates the location of the patient by tracking up to 22 satellites with a horizontal position accuracy of 2.5 meters. Once the module has a GPS lock, the microcontroller uses serial communication to communicate with the GPS module and receives the patient's location coordinates in latitude (N/S) and longitude (E/W).

These latitude and longitude coordinates are sent to the medical personnel's mobile phone through simple text messages using a GSM module. The microcontroller is programmed to send these texts with the most recent coordinates from the GPS module to an already saved contact number of the medical personnel after every two minutes. The GSM module works on the GSM network using any 2G/3G/4G SIM card on a global network, hence it provides a wide network coverage and it's possible to send simple text messages even if there is no internet connectivity in certain areas. Thus, the patient's updated and most recent location is sent to the medical personnel after every two minutes so that in case the patient having an emergency is in a moving state, the EMT/doctor will receive real-time updates on the changes in location of the patient and navigate to the last updated location from the most recent text message.

The text messages are provided with a hyperlink of the patient's location, which directly opens Google Maps in the medical personnel's mobile phone, and they can navigate to the patient's location immediately. Thus, this fail-safe model's operation will provide the real-time location updates of the patient even if the primary operation model of the wearable device faces any issue in performing its task, and the patient will receive immediate medical assistance that may, in turn, save their life.

An ECG is the curve for measuring the overtime shift in the cardiac electrical potential. We can detect bioelectric signals by placing the electrode terminals at particular strategic spots on the skin of the human body. These strategic spots help us extract the most accurate data available. Electrode communication mode is also known as the lead method. Currently, the twelve-lead standard system is the lead device most commonly used for medical use. Ten electrodes must be placed on different positions on the human body, and eight-channel data are measured simultaneously: two leads on the limb and six leads on the chest. A twelve-lead ECG system has the potential to extract the cardiac electrical activity of a subject completely.

The number of electrodes, however, is very large, and the portability and accessibility are relatively poor, so it is mostly used at the hospital for static ECG inspection. Our system answers to these problem statements of portability and accessibility by reducing the number of electrodes to three by using the three-electrode placement system rather than the traditional twelve-lead system. The data is measured at three channels: two limbs leads and a lead in the chest. According to the ECG detection theory, chest leads are unipolar leads, hence there is no change in the mode of acquisition of data from the chest leads. So only one lead electrode in the chest is used to replace six leads in the chest. All chest leads data can be identified by changing electrodeposition in different periods.

2.3 SOFTWARE ARCHITECTURE OF WEARABLE USER-END DEVICE

2.3.1 ARCHITECTURE OF SOFTWARE AT FRONT END

The architecture of the system software at the user end can be broken down into two primary software sections: driver layer and application layer as shown in Figure 2.2. The driver layer consists of the initiating program for the microcontroller and interfacing the microcontroller with external sensors and data transfer modules. The application layer controls the operation of AD8232 for capturing the ECG signal, data transfer using wireless Bluetooth technology and data storage, etc.

Sensing and capturing an accurate ECG signal from the patient is one of the most crucial functions of the system. An ECG is a graphical representation of the electrical activity of the heart over a period of time, which is recorded by the electrodes connected to the body either using three leads or twelve leads attached to the surface of the skin. To reduce bulkiness and keeping user comfort in mind, we have used a three-lead ECG system. The ECG signal is corrupted by different noises, which can be divided into two main categories: high-frequency noises and low-frequency noises. These noises lead to wrong interpretations of the ECG signal.

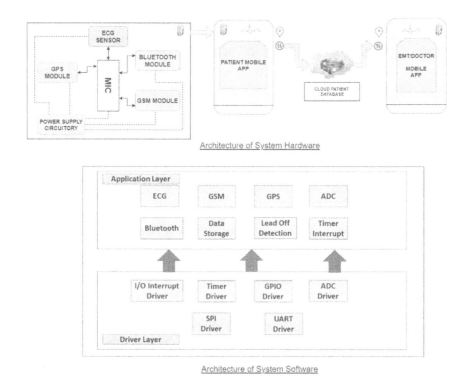

FIGURE 2.2   Architecture of System Hardware and Software.

These interferences can be due to power line interference, channel noise, baseline wander, electromyogram (EMG) noise, electrode contact noise, and motion artifacts.

2.3.1.1   Algorithm for Measuring ECG Signal

This program will perform the ECG signal measurement of the patient and send the patient's ECG signal data to the microcontroller to perform further operations on this data.

> **Step 1:** In the void setup (), Initialize serial communication at a specified baud rate of 9,600. Specify two GPIO pins on the microcontroller that act as input pins to detect if the ECG leads are off/disconnected.
> **Step 2:** Check for high output at either of the assigned input pins. If true, print the message "ECG Leads Disconnected!" Else, go to step 3.
> **Step 3:** Read the ECG data through the analog pin A0 and print/plot this data on serial monitor/plotter.
> **Step 4:** Set a delay for 20 microseconds and then go to step 2.

Calculating the accurate location of the patient as quickly as possible is the other most crucial function of the Zentraq device system and can save countless lives if provided to medical personnel like EMTs and doctors. There are two systems in the device that calculate and provide the patient's location. First, the GPS location module onboard the wearable device and second, the mobile phone app's location feature.

During the preliminary design of the system, we used the SIM7600EI 4G/GSM/GPRS/GPS UART modem. The reason for selecting this modem was that it provided both GPS service as well as internet connectivity. Both of these are key requirements for the Zentraq device to function properly. But this modem had some practical problems that were discovered during the testing phase of the device. The modem being a hybrid component providing multiple functionalities drew a large amount of current (4.1–4.3 A) from the power supply, making it difficult to incorporate in a wearable device. Also, the accuracy of the location calculated by the GPS module decreased after using it for a longer duration on a battery.

To solve these problems, we opted for the u-blox NEO-6M GPS module. It proved to be a much better fit for our requirements due to its compact size, accurate location calculation, and low operating voltage and current parameters. The Zentraq mobile phone app uses the phone's inbuilt GPS location services to calculate the accurate location of the patient and provide a Google Maps link that can directly navigate to the patient's location. This works very accurately in an area with proper internet connection, but if the user is not in an area with proper connectivity, this system fails to provide the exact location of the patient. This is where the fail-safe operation module of the wearable device proves to be crucial and lifesaving by providing the location of the patient through a simple text message that can be sent over a 2G/3G connectivity that is available in almost all parts of the world using the onboard SIM800L GSM/GPRS module.

2.3.1.2 Fail-Safe Mechanism Algorithm

This program will allow us to calculate the location of the patient and send it to the medical personnel through simple text messages.

Tiny GPS++ and SoftwareSerial libraries are included for setting up a serial connection and communicating with GPS and GSM modules. Tiny GPS++ is a small GPS library used for Arduino providing basic National Marine Electronics Association (NMEA) parsing. NMEA is a standard data format supported by all GPS manufacturers.

Step 1: Include TinyGPS++ and SoftwareSerial libraries and set the baud rate to 9,600.

Step 2: Create SoftwareSerial objects for communication with GPS and GSM modules.

Step 3: In the void setup(), begin serial communication at a specified baud rate of 9,600.

Step 4: Check whether the data is available on the GPS serial port else go to step 8.

Step 5: If the data is available, then read the data from the GPS serial port.

Step 6: Check whether the available data is valid.

Step 7: If the location data is valid, extract the latitude (N/S) and longitude (E/W) and send the data through a simple text message to the medical personnel using the GSM module and go to step 9.

Step 8: If there is no incoming data from the serial port for more than five seconds then, print "NO GPS Detected."

Step 9: Set a delay for two minutes and then go to step 4.

The fail-safe mechanism is tested accordingly and results are obtained on the Serial Plotter and the mobile phone.

Once the location and ECG data are collected from the GPS module and ECG sensor that are onboard the wearable device, the ECG data needs to be transferred to the patient's mobile phone to further upload it to the Cloud Patient Database. This data transfer needs to be done wirelessly to make the system more hassle-free and user-friendly. There were two wireless standards to consider for wireless data transfer: Wi-Fi technology, and Bluetooth technology. For our system requirements, Bluetooth is a better option because it is suitable for the device-to-device connection within a shorter range of communication, unlike Wi-Fi, which is better suited for operating large networks with multiple device connections spread over a larger area. Another advantage of using Bluetooth technology is that it consumes less power than Wi-Fi technology, which makes it more suitable for our battery-operated wearable device.

We are using the HC-05 Bluetooth SPP module, which is designed for transparent wireless serial connection setup and communication. As HC-05 uses serial communication (USART) it can be easily interfaced with any controller or PC with a programmable baud rate and supports the following baud rates: 9,600, 19,200, 38,400, 57,600, 115,200, 230,400, 460,800 bps.

Although there are a few precautions that must be followed to ensure proper connection and data transfer with no or negligible data loss, the module requires a steady power supply within its operation range and works perfectly fine on a 3.3 V VCC supply. A constant supply of higher voltage may cause permanent damage to the module and hence must be avoided. Also, the HC-05 Bluetooth module operates on a 3.3 V level for serial communication, hence a resistive circuitry is used to drop the voltage of the signal coming from the microcontroller. If this provision is not made, the data transfer between the HC-05 module and the mobile phone is not successful and garbage data is received at the receiver's end.

2.3.1.3 Bluetooth Data Transfer Algorithm

This program will allow the transfer of ECG signal data from the hardware setup onboard the wearable device to the patient's mobile phone.

The SoftwareSerial library is included for setting up a serial connection for data transfer communication between the microcontroller and the HC-05 Bluetooth module.

Step 1: Include the SoftwareSerial library and set the baud rate to 9,600.

Step 2: Create a SoftwareSerial object for communication with the HC-05 Bluetooth module.

Step 3: In the void setup(), Begin serial communication at a specified baud rate of 9,600. Specify two GPIO pins on the microcontroller that act as input pins to detect if the ECG leads are OFF.

Step 4: Check for high output at either of the assigned input pins. If true, print the message "ECG Leads Disconnected!" Else go to Step 5.

Step 5: Read the ECG data through the analog pin A0 and, send this data to the Bluetooth module using the created SoftwareSerial object for further transmission to the user patient's mobile phone.

Step 6: Set a delay for one microsecond and then go to step 4.

2.3.2 ANDROID APPLICATION FOR LOCATION EXTRACTION AND DATA TRANSMISSION

The Zentraq system has two mobile phone apps specifically designed to perform a certain crucial operation in the functioning of the system. These two apps are designed for the patient and the EMT/doctor (medical personnel) respectively.

The patient app's function is receiving the patient's ECG data from the wearable device through a Bluetooth connection and upload this data to the patient data storage server using the mobile phone's internet connection or a Wi-Fi connection. Along with this, it collects the patient's location using the mobile phone's GPS location service and uploads this information to the patient data storage server similarly to the patient's ECG data.

After successfully installing the app on the patient's mobile phone, the app is opened and it asks the user permission to access the device's location and Bluetooth. Once these permissions are granted, the app asks the user permission to enable and use their mobile phone's GPS. After all the permissions are granted by the user, the Bluetooth connection between the patient's mobile phone and the HC-05 Bluetooth module onboard the wearable device is established successfully. Following the successful connection, ECG data transfer from the wearable device to the patient's mobile phone is started. After preliminary testing, it was observed that in some cases few data packets were corrupted during transmission from the hardware setup to the patient's mobile phone. This corrupted data further leads to problems in plotting the ECG report of the patient, hence a provision is made to filter out corrupted data in this patient app during the data transmission itself. Once the ECG data transfer is completed, this data along with the location of the patient calculated using the mobile phone's GPS is uploaded to the patient data storage server immediately. This app acts as an important transmitting link in the data transmission process of the Zentraq system.

The second mobile app of the system is designed for medical personnel's use. The purpose of this app is to function as a receiving link in the data transmission process. It requires access to the user's device location. This app downloads the patient's location and ECG signal data from the Cloud Patient Database into the user medical personnel's mobile phone. This downloaded data is then separated and displayed in different formats. The location of the patient having a medical emergency can be accessed directly in Google Maps and the medical personnel can quickly navigate to the patient's location without wasting time, which can save the patient's life.

The patient's ECG signal data is presented graphically in the app by plotting the ECG data in a Serial Plotter to form an ECG report of the patient as shown in Figure 2.3. This report can be examined properly by the medical personnel by scrolling sideways through the ECG reports forming a preliminary diagnosis that will help

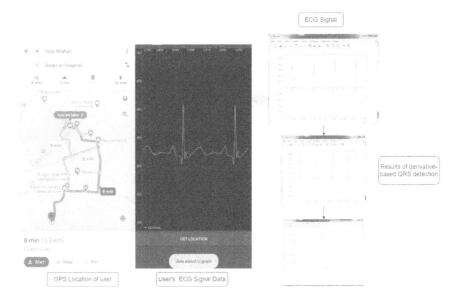

FIGURE 2.3   Results and Validation.

them prepare the necessary equipment and arrange necessary medications for the patient. All of this lifesaving information will help in saving precious time during the golden hours and may save the patient's life.

2.4   VALIDATION OF SYSTEM

We carried out rigorous testing of our system in various situations to test how all the functions and features operated under emergency circumstances. For validation of the system's operation, we have tested it on 3 test subjects, which include middle-aged adults as well as elderly patients. The outcomes of the testing are summarized next.

2.4.1   Speed of System Operation (Start to End)/System Latency

The system's speed can be defined as the overall time duration that elapses from the moment the patient presses the SOS button on the wearable device until the medical personnel receive the patient information. We conducted five tests to find out the system's latency and averaged the outcomes to find an approximate value of the time duration. There are two system operation speeds once the SOS button is pressed:

- The time duration is taken for generation and delivery of the simple text messages containing the patient's location updates to the medical personnel phone number. This came out to be around 16 seconds.
- The time duration is taken for the patient's ECG to be recorded and getting uploaded to the Cloud Patient Database along with the GPS location data of the patient's mobile phone. The ECG data recording and transfer via Bluetooth take approximately 36 to 40 seconds. This data is immediately

uploaded to the Cloud Patient Database using the Zentraq patient app. It takes around 4 to 5 seconds for the ECG report and patient location to be updated and made available for use on the medical personnel's app. Thus, the complete operation of the system is completed within 40 to 60 seconds ideally.

2.4.2   PATIENT LOCATION ACCURACY

The system provides accurate patient location leading to quick navigation and saves precious time during the golden hour, which is crucial in saving the patient's life. For fast and error-free navigation, a Google Maps link of the patient's location is provided as it uses the shortest possible routes while navigating a user to their destination. The accuracy of location depends on the position of the patient:

- If the patient presses the SOS button while indoors, in an enclosed environment, it takes a little longer for the device to obtain a GPS lock on the patient's location, and the accuracy is around 10 to 50 meters horizontally.
- If the patient presses the SOS button while outdoors, the device obtains a GPS lock in a few seconds, and the accuracy of this location is up to 10 meters horizontally.

2.4.3   ECG REPORT ANALYSIS

The ECG signal that is captured by the wearable device is analyzed and processed to determine its authenticity and establish its medical accurateness.

2.4.3.1   Doctor's Analysis (Signal Structure and Quality Comparison with Medical-Grade ECG Report)

We consulted with a cardiologist for comparing our ECG reports with actual medical-grade ECG reports, and the doctor commented that the morphological structure of our ECG plot is similar to that of an actual ECG plot obtained in a medical-grade ECG report, and a preliminary diagnosis about the patient medical condition can be formed.

2.4.3.2   ECG Signal Processing and Peak Detection

We used the QRS-complex detection operation [29,30] for the ECG analysis. Detection of the QRS complex is one of the most substantial tasks in automated ECG signal analysis. The ECG signal has a nonstationary nature and complex shape that changes from one human to the other, making the QRS-complex detection a complicated task. We used a simple derivative-based QRS detection algorithm.

2.4.3.3   Simple Derivative-Based QRS Detection Algorithm

Step 1: Signal filtering (LF, HF)
Step 2: Calculation of the first and the second derivatives

$$y'(n) = x(n) - x(n-2)$$

$$y''(n) = x(n) - 2x(n-2) + x(n-4),$$

Where (n) is the input ECG signal, $y'(n)$ is the first derivative and $y''(n)$ is the second derivative of the input signal.

Step 3: Resulting combination

$$res = 1,3 \cdot \left| y' \right| + 1,1 \cdot \left| y'' \right|$$

Step 4: Choosing the threshold A
Step 5: Close peaks removed using the time limit between QRS complexes.
Step 6: Maximum calculation for R-peak detection.
Step 7: Zero-crossing event detection to find the end of the R-peak.

Thus, we were able to extract and display distinct features from the ECG signal captured by our device. These features are used by doctors in analyzing and forming a diagnosis of a patient's medical condition.

2.5 CONCLUSION

In this chapter, we have presented a smart, ready-to-use, sensor system that would help in acquiring and transmitting the ECG and location of a person at a crucial time. This data is shown to have vastly helped EMS to bring down the mortality rate, as well as give the doctors a better understanding of the cause and effect of the trauma in many countries. Countries such as India would immensely benefit if a device, such as the one proposed in this chapter, were to go to market. With a rising number of fatalities due to an untimely response from respective authorities, there is a dire need for home care health accessories, and Zentraq exactly meets this requirement. The goal is to increase the accuracy of monitoring and sensing technology. Wearing electrode patches that come with the system may limit patient movement, but in our system, leads can be placed at different spots, as per the requirement, and there is also the flexibility of using different kinds of electrodes, as they are clipped onto one of the sockets, thus enabling a more discrete and comfortable wearing of the system. Preliminary test results have shown that Zentraq has the potential to be integrated into wearable technologies in rehabilitation technologies. This wearable medical healthcare device is proposed as a solution for the increasing need for automated collection of health data from multiple patients both inside and outside of a medical environment (hospital or nursing home).

ACKNOWLEDGMENT

We would like to express our deep gratitude to Dr. Aniket Joshi (MBBS and FCPS), senior doctor, and intensivist at KEM Hospital, Pune, for his patient guidance, enthusiastic encouragement, and useful critiques of this research work.

REFERENCES

1. L. Gatzoulis and I. Iakovidis, "Wearable and Portable eHealth Systems," *IEEE Engineering in Medicine and Biology Magazine* 26, no. 5 (September–October 2007): 51–56.

2. A. Lymperis and A. Dittmar, "Advanced Wearable Health Systems and Applications, Research and Development Efforts in the European Union," *IEEE Engineering in Medicine and Biology Magazine* 26, no. 3 (May/June 2007): 29–33.
3. G. Tröster, "The Agenda of Wearable Healthcare," *IMIA Yearbook of Medical Informatics* (Stuttgart, Germany: Schattauer, 2005), 125–38.
4. "Global Data," *The Economic Times*, https://economictimes.indiatimes.com/topic/GlobalData. (August 2020)
5. Y. Hao and R. Foster, "Wireless Body Sensor Networks for Health-Monitoring Applications," *Physiological Measurement* 29, (November 2008): R27–R56.
6. United Nations, Department of Economic and Social Affairs (Population Division), *World Population Ageing 2013* (New York: United Nations, 2013).
7. C. O'Shaughnessy, "The Basics: National Spending for Long-Term Services and Supports: 2012, National Health Policy Forum," (National Health Policy Forum, George Washington University, Washington, DC, 2013).
8. M. S. Sidhu, L. Griffith, K. Jolly, P. Gill, T. Marshall, and N. K. Gale, "Long-Term Conditions, Self-Management and Systems of Support: An Exploration of Health Beliefs and Practices within the Sikh Community," *Ethnicity and Health* 21, no. 5 (October 2016): 1–17.
9. World Health Organization. "Cardiovascular Diseases (CVDs)," accessed May 17, 2017, https://www.who.int/health-topics/cardiovascular-diseases#tab=tab_1.
10. S. Park and S. Jayaraman, "Enhancing the Quality of Life Through Wearable Technology," *IEEE Engineering in Medicine and Biology Magazine* 22, no. 3 (2003): 41–48. https://doi.org/10.1109/memb.2003.1213625.
11. V. Jones et al., "Remote Monitoring for Healthcare and for Safety in Extreme Environments," in *M-Health Topics in Biomedical Engineering*, eds. R. S. H. Istepanian, S. Laxminarayan, and C. S. Pattichis (Boston: Springer, 2006), https://doi.org/10.1007/0-387-26559-7_43.
12. A. Wolfe, "Institute of Medicine Report: Crossing the Quality Chasm: A New Health Care System for the 21st Century," *Policy, Politics, & Nursing Practice* 2, no. 3 (August 2001): 233–35. https://doi.org/10.1177/152715440100200312.
13. H. Ren, M. Q.-H. Meng, and X. Chen, "Physiological Information Acquisition Through Wireless Biomedical Sensor Networks," (presentation, 2005 IEEE International Conference on Information Acquisition, Hong Kong, China, 2005), https://doi.org/10.1109/ICIA.2005.1635137.
14. A. Caduff, E. Hirt, Y. Feldman, Z. Ali, and L. Heinemann, "First Human Experiments with a Novel Non-Invasive, Non-Optical Continuous Glucose Monitoring System," *Biosens Bioelectron* 19, no. 3 (November 2003): 209–17, https://doi.org/10.1016/s0956-5663(03)00196-9.
15. A. Pantelopoulos and N. G. Bourbakis, "A Survey on Wearable Sensor-Based Systems for Health Monitoring and Prognosis," *IEEE Transactions on Systems, Man, and Cybernetics,* Part C (Applications and Reviews) 40, no. 1 (January 2010): 1–12, https://doi.org/10.1109/TSMCC.2009.2032660.
16. J. Muhlsteff, O. Such, R. Schmidt, M. Perkuhn, H. Reiter, J. Lauter, and M. Harris, "Wearable Approach for Continuous ECG - and activity patient-monitoring." (presentation, The 26th Annual International Conference of the IEEE Engineering in Medicine and Biology Society, San Francisco, California, 2004), https://doi.org/10.1109/iembs.2004.1403638.
17. N. Oliver and F. Flores-Mangas, "HealthGear: A Real-Time Wearable System for Monitoring and Analyzing Physiological Signals." *International Workshop on Wearable and Implantable Body Sensor Networks (BSN'06)*: 4–64, https://doi.org/10.1109/bsn.2006.27.
18. S. Guillen, M. T. Arredondo, V. Traver, J. M. Garcia, and C. Fernandez, "Multimedia Telehomecare System Using Standard TV Set," *IEEE Transactions on Biomedical*

Engineering 49, no. 12 (December 2002): 1431–37, https://doi.org/10.1109/TBME.2002.805457.

19. J. Muhlsteff et al., "Wearable Approach for Continuous ECG - and Activity Patient-Monitoring," (presentation, The 26th Annual International Conference of the IEEE Engineering in Medicine and Biology Society, San Francisco, CA, 2004): 2184–2187, https://doi.org/10.1109/IEMBS.2004.1403638.

20. E. A. Miller, "Telemedicine and Doctor-Patient Communication: An Analytical Survey of the Literature," *Journal of Telemedicine and Telecare* 7, no. 1 (2001): 1–17, https://doi.org/10.1258/1357633011936075.

21. P. Bonato, "Advances in Wearable Technology and Applications in Physical Medicine and Rehabilitation," *Journal of NeuroEngineering and Rehabilitation* 2 (February 2005): 2.

22. S. Eaton, S. Roberts, and B. Turner, "Delivering Person Centered Care in Long Term Conditions," *British Medical Journal* 350 (February 2015): h181, https://doi.org/10.1136/bmj.h181.

23. P. J. Soh, G. A. E. Vandenbosch, M. Mercuri and D. M. M. Schreurs, "Wearable Wireless Health Monitoring: Current Developments, Challenges, and Future Trends," *IEEE Microwave Magazine* 16, no. 4 (May 2015): 55–70.

24. Arduino.cc, "Arduino UNO documentation," https://store.arduino.cc/usa/arduino-uno-rev3. (August 2020)

25. Components101, "HC-05 Bluetooth Module Datasheet," https://components101.com/wireless/hc-05-bluetooth-module. (August 2020)

26. Analog Devices, "AD8232 Datasheet," https://www.analog.com/media/en/technical-documentation/data-sheets/ad8232.pdf. (August 2020)

27. P. Bonato, "Wearable Sensors/Systems and Their Impact on Biomedical Engineering," *IEEE Engineering in Medicine and Biology Magazine* 22 (May–June 2003): 18–20.

28. u-blox, "NEO-6 GPS module datasheet," https://www.u-blox.com/sites/default/files/products/documents/NEO-6_DataSheet_%28GPS.G6-HW-09005%29.pdf. (August 2020)

29. S. Hamdi, A. Ben Abdallah, and M. H. Bedoui, "Real Time QRS Complex Detection Using DFA and Regular Grammar," *BioMedical Engineering OnLine* 16, no. 31 (2017), https://doi.org/10.1186/s12938-017-0322-2. (August 2020)

30. B. Kohler, C. Hennig, and R. Orglmeister, "The Principles of Software QRS Detection," *IEEE Engineering in Medicine and Biology Magazine* 21, no. 1 (January–February 2002): 42–57, https://doi.org/10.1109/51.993193.

3 AI and Deep Learning for Biomedical Image Analysis

R. Priyatharshini and R. Reena Roy

CONTENTS

3.1 INTRODUCTION

In recent years, deep learning (DL) and artificial intelligence (AI) have experienced enormous growth in the medical field. Algorithms of DL and AI are playing a vital role in healthcare such as image processing with scan images, building an efficient computer-aided diagnosis system for image interpretation, image segmentation, and classification. Machine learning (ML) techniques mine exact information from the scan images and interprets information accurately. AI and DL algorithms assist doctors in making better decisions for diagnosis and help in assessing the severity of various diseases and prevent the spread in time. These techniques comprised of various unsupervised algorithms like support-vector machine, neural network, k-nearest neighbor, etc. and DL algorithms such as convolutional neural networks, recurrent neural network, long short-term memory, extreme learning model etc. [1]. Although early detection of various diseases with existing computer-aided diagnosis systems in medical imaging have shown substantial accuracies, but advancements in ML techniques have an exponential growth in the domain of deep learning [2]. DL techniques showed an encouraging performance in various fields like speech recognition, text recognition, and image processing.

The main goal of this chapter is to give an overview of DL algorithms and AI in medical image analysis in terms of current and future research scope. This chapter provides in-depth knowledge about algorithms of DL in the domain of healthcare.

3.2   HEALTHCARE DATA SOURCES

3.2.1   ELECTRONIC HEALTH RECORDS

Electronic health records (EHRs) are actual data that make patients' information instantly available for legal users. An EHR is a piece of software used to document, store, and analyze a patient's data more securely. It can be accessed either locally or remotely by authenticated users. EHRs may also contain diagnosis information, treatment planning, and radiology, laboratory and test results.

3.2.2   BIOMEDICAL IMAGES

Biomedical images are obtained through medical imaging, and the images are used for diagnostic purposes. The biomedical images are captured by various technologies such as X-ray, PET scan, MRI, and CT scan, which help screen various organs and tissues. DL algorithms can provide in-depth analysis to detect patterns and characteristics indicative of tumors and other ailments. DL algorithms are widely used in healthcare to detect diseases as early as possible by training the neural networks and disease patterns can be identified easily. Imaging modalities such as computed tomography (CT), magnetic resonance imaging (MRI), and positron emission tomography (PET) are most commonly used to detect the disease patterns to retrieve meaningful information.

3.2.3   SENSOR DATA IN HEALTHCARE

Sensors are widely used in the healthcare industry to collect signals from the human body and convert them into electrical signals. The various forms of sensors used in healthcare are in CT scan, PET scan, MRI scan, ECG, etc. An oximeter is a device that acts as a sensor for collecting the level of oxygen in the blood and the pulse rate. These sensors are used for collecting data that plays a big role in the healthcare industry. These sensors are particularly useful for patients with high risk factors and also serve as a lifesaver by generating data that is useful for clinicians to make better decisions. This device is impacting the healthcare industry in numerous ways, and it serves as a monitoring tool for disabled patients.

3.2.4   BIOMEDICAL SIGNAL ANALYSIS

The human body frequently communicates health condition information. This information will be collected through sensors that generate electrical signals, which help in measuring the pulse rate, heart rate, level of oxygen in the blood, blood glucose for diabetic patients, nerve conditions for monitoring the activity of the brain, and so on. Usually, these signal measurements are taken at regular intervals of time and updated on a patient's record. Doctors keep tracking these records and make treatment decisions and plans for surgery. Signal processing domain based on clinical field gives a measurable analysis of data to provide useful information upon which the doctors can use to make effective decisions. Researchers are determining new pathways to process the biomedical signals using statistical techniques and DL algorithms. Based on the statistical tools, the signals generated will be computed by software that provides inputs to clinicians with real-world information and provides greater insights to aid in clinical assessments and surgical planning.

3.2.5 GENOMIC DATA

Genomics is a field of biology that mainly deals with the structure, function, and activity of genomes. From a technical point of view, a genome is a human complete set of sequenced DNA. This genomic data is one of the sources of data for healthcare. The genome of a person reveals the susceptibility of diseases and gives information about the people who are closely related to them.

3.3 DEEP LEARNING FOR HEALTHCARE

DL is a subclass of ML, having types of artificial neural networks (ANNs) that look like multilayered human reasoning systems. DL is currently highly successful in working with research projects mainly because of its ability to process clinical data. The ANN was introduced in 1950, but there were certain challenges that still existed due to a lack of training in network architecture and overfitting problems, a lack of computing power, and insufficient data to train the neural network. Many challenges have been overcome due to enriched computation capacity with graphical processing units (GPUs) with the current accessibility of big data and new evolving techniques to train a deep neural network (DNN). These DL techniques have achieved remarkable outcomes in simulating human beings in various domains like medical imaging. DL algorithms are used widely in the field of medical image processing by providing better results with accuracy in various applications like segmentation and classification. These DL approaches can be used for the segmentation (removing some portion) of the lungs for detecting tumors and other structures in the brain like biological cells and tissues, pancreatic tumors, cardiac stenosis, etc. All these previously mentioned applications widely take intensity and spatial consistency as inputs, and computations are fed into neural networks such as various graphical models. The usage of DL and AI in medical imaging is very well adopted by radiologists for the interpretation of results. One of the most important criteria for the advancement of AI and its implementation is done by both doctors and engineers. With the recent scientific advancements by ImageNet, very large and fully marked-up databases are needed in the field of medical imaging. This will be vibrant to train the neural network to improve the system throughput.

3.4   RESEARCH CHALLENGES IN BIOMEDICAL IMAGE ANALYSIS

There are various research challenges in biomedical image analysis. In recent years, biomedical imaging systems have produced tremendous amount of images that contain large amounts of enriched information. The enriched information is hidden in the data, and image processing techniques are required to extract the valuable information for clinical decisions by doctors. Rapid growth in advancements of medical image analysis over the years reveals that there are as many ideas available to provide solutions to real-world problems with enough throughput relative to precision, trustworthiness, and speed. Several new challenges have arisen. First, newer image processing technologies are required that can be proficiently utilized for the exact same medical task. Second, the original images, that is, ground truth, are required for validation and ML tasks. Third, techniques for analyzing heterogeneous image sources are needed. Finally, the internal structure and parts of the body models have a greater impact in many applications, and methodologies to construct models from medical images are required. These limitations need to be overcome by implementing reliable and faster algorithms as solutions to real-world problems.

3.5   DEEP LEARNING ALGORITHMS FOR BIOMEDICAL IMAGE ANALYSIS

3.5.1   MELANOMA SKIN DISEASE DIAGNOSIS

Dermoscopy basically allows the visualization of diagnostic features of pigmented skin lesions. This technique is useful for dermatologists in distinguishing benign from malignant lesions. The real-time monitoring of the disease is not done using this system. It only helps to diagnose the disease at some predicted time by medical advisors. An algorithmic technique called SVM came into existence because it worked really well with a clear margin of separation, but the drawback is that since the data set is large, it does not perform well because the training time is higher. Since the dermoscopy image data set are available only in small amounts, it hinders the training of networks for the automated diagnosis of pigmented skin lesions [3]. The dataset from HAM10000 ("human against machine with 10,000 training images") contains 10,000 images of pigmented lesions and melanoma images with the subsequent layer mask present for feature extraction [4]. Keeping these issues in mind, this chapter proposes a new automatic method for discrimination between pigmented lesions and skin cancer and also real-time monitoring using the latest generation technique called convolution neural network (CNN), which acts far better than the already existing support vector machines (SVMs) algorithm on a big data set with a short period of training time. The main benefit of using CNN is that it is capable of representing visual things very impressively depending on the training dataset for the recognition tasks. Figure 3.1 shows the step-by-step process involved in skin disease classification.

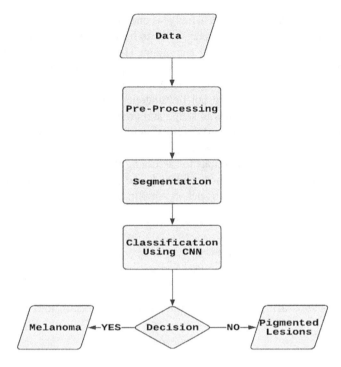

FIGURE 3.1   Step-by-step process involved in skin disease classification.

3.5.2 PANCREATIC DISEASE DIAGNOSIS

Pancreatic cancer is one of the deadly diseases and accounts for a huge death rate around the globe [5]. Early detection of pancreatic cancer involves screening of patients through scans such as CT, MRI, PET, X-ray, and U/S [6]. Automatic segmentation of the pancreas from abdomen CT images is the basic issue in biomedical image processing and computer-aided diagnosis (CAD) [7]. Meanwhile segmentation of the pancreas is a very challenging task because of its location, which is situated behind the stomach and internal structure of the other organs [8]. Due to the exponential growth in the CAD system, the necessity for clinical diagnosis, segmenting a required portion of the image from the background is highly demanded [9]. For early detection of pancreatic cancer, automatic segmentation of the pancreas from the abdomen in CT scan images is widely used by the healthcare industry [10]. This is driven by both AI and DL algorithms for classifying cancer as benign and malignant. AI [11, 12] reveals studies on the deep supervision techniques on pancreatitis and pancreatic cancer detection by using computing techniques.

3.5.3 LUNG DISEASE DIAGNOSIS

Diagnosis of lung disease from lung CTs includes processing the image before detecting cancerous lesions and also includes segmentation [13]. Removing the required

ROI (Region of Interest) from the CT image is one of the challenging tasks, since the images are irregular in shape, complex, and can have incomplete values (noisy) that affect the accuracy of segmentation techniques [14]. Early detection of malignant cells from CT images is a difficult task, with the present segmentation techniques such as region growing methods, etc. [15]. A lot of research is still being conducted, but an efficient framework could not be modelled since the images were having some issues such as irregularities in shape and noisy values [16]. Automatic detection of tumors from CTs for an early prediction of cancer is a tremendous research area in the field of AI [17]. This work can be exploited for periodic monitoring of the individual patient's health conditions for assessing the risks [18]. CAD systems are widely used as a tool for clinicians to identify the presence of tumors in individuals with the following risk factors such as smoking, obesity, rashes, family history, frequent exposure to pollutants, etc. [19].

3.5.4 CARDIAC DISEASE DIAGNOSIS

Cardiovascular diseases are a group of disorders that end up in diseases like stroke and tightening the vessels of the blood by forming a fatty substance called plaque, which causes an attack [20]. Due to plaque, a blood clot occurs and as a result, the

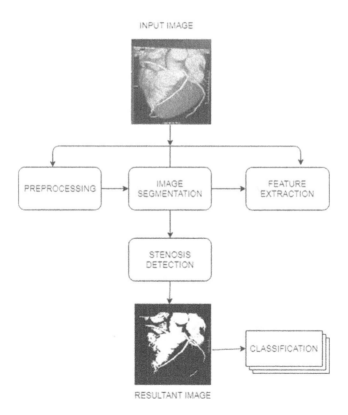

FIGURE 3.2   Process of coronary artery stenosis detection and classification.

coronary artery becomes very narrow. Because of its risk, it causes around a third of the world's deaths, so early detection is essential through a computer-aided diagnosis system [21]. The skillful approach to examine the coronary artery blockage is angiography. Due to the existing challenges in cardiovascular diseases, severity of stenosis is required in the early stage [22]. This work proposes an image processing method (segmentation) for localizing the stenosis region in the coronary artery [23]. This proposed method takes a set of CT images as input and produces much accuracy and better segmentation results, which help physicians better diagnose problems and plan better for surgery [24]. Various segmentation algorithms are incorporated to localize the regions in the artery. This method is applied on a set of reconstructed cardiac CT images along with some preprocessing for the detection of blockage. Figure 3.2 shows the process of coronary artery stenosis detection and classification.

3.6   CONCLUSION AND FUTURE WORK

This chapter gives an in-depth analysis of the AI and DL algorithms in the domain of medicine. Usage of DL algorithms helps in recognizing the complex patterns from biomedical images. DL techniques help in processing healthcare data, which results in early detection of disease patterns, personalized optimal treatment plans, and a fast and more accurate diagnosis. This chapter summarizes the overview of DL algorithms and AI for biomedical image analysis.

REFERENCES

1. Liang, M., & Hu, X. (2015). Recurrent convolutional neural network for object recognition. In *Proceedings of the IEEE conference on computer vision and pattern recognition* (pp. 3367–3375).
2. De Mulder, W., Bethard, S., & Moens, M. F. (2015). A survey on the application of recurrent neural networks to statistical language modeling. *Computer Speech & Language, 30*(1), 61–98.
3. Ali, A. R. A., & Deserno, T. M. (2012, February). A systematic review of automated melanoma detection in dermatoscopic images and its ground truth data. In *Medical Imaging 2012: Image Perception, Observer Performance, and Technology Assessment* (Vol. 8318, p. 83181I). International Society for Optics and Photonics.
4. Tschandl, P., Rosendahl, C., & Kittler, H. (2018). The HAM10000 dataset, a large collection of multi-source dermatoscopic images of common pigmented skin lesions. *Scientific Data, 5*, 180161.
5. Dimastromatteo, J., Brentnall, T., & Kelly, K. A. (2017). Imaging in pancreatic disease. *Nature reviews Gastroenterology & Hepatology, 14*(2), 97.
6. Al Mamun, K. A., & McFarlane, N. (2016). Integrated real time bowel sound detector for artificial pancreas systems. *Sensing and Bio-Sensing Research, 7*, 84–89.
7. Kenner, B. J., Go, V. L. W., Chari, S. T., Goldberg, A. E., & Rothschild, L. J. (2017). Early detection of pancreatic cancer: the role of industry in the development of biomarkers. *Pancreas, 46*(10), 1238.
8. Sakamoto, H., Harada, S., Nishioka, N., Maeda, K., Kurihara, T., Sakamoto, T., ... & Kudo, M. (2017). A social program for the early detection of pancreatic cancer: the Kishiwada Katsuragi Project. *Oncology, 93*(Suppl. 1), 89–97.

9. McGuigan, A., Kelly, P., Turkington, R. C., Jones, C., Coleman, H. G., & McCain, R. S. (2018). Pancreatic cancer: A review of clinical diagnosis, epidemiology, treatment and outcomes. *World Journal of Gastroenterology, 24*(43), 4846.

10. Miller, R. J., Han, A., Erdman Jr, J. W., Wallig, M. A., & O'Brien Jr, W. D. (2019). Quantitative ultrasound and the pancreas: Demonstration of early detection capability. *Journal of Ultrasound in Medicine, 38*(8), 2093–2102.

11. Kaissis, G., & Braren, R. (2019). Pancreatic cancer detection and characterization—state of the art cross-sectional imaging and imaging data analysis. *Translational Gastroenterology and Hepatology, 4.*

12. Zhou, Y., Xie, L., Fishman, E. K., & Yuille, A. L. (2017, September). Deep supervision for pancreatic cyst segmentation in abdominal CT scans. In *International conference on medical image computing and computer-assisted intervention* (pp. 222–230). Springer, Cham.

13. Damodar Dipika A, & Prof. Krunal Panchal. (December 2015). A survey of lung tumor detection on CT images. *JETIR* (ISSN-2349-5162), 2(12).

14. Zhang, D. Q., & Chen, S. C. (2004). A novel kernelized fuzzy c-means algorithm with application in medical image segmentation. *Artificial Intelligence in Medicine, 32*(1), 37–50.

15. Deepthy Janardhanan, & Dr.R.Satishkumar. (March 2017). Lung cancer detection using active contour segmentation model. *International Journal of Innovative Research in Advanced Engineering (IJIRAE)* (ISSN: 2349–2163),4(03, Special Issue).

16. Ji, H., He, J., Yang, X., Deklerck, R., & Cornelis, J. (2013). ACM-based automatic liver segmentation from 3-D CT images by combining multiple atlases and improved mean-shift techniques. *IEEE Journal of Biomedical and Health Informatics, 17*(3), 690–698.

17. Lai, J., & Ye, M. (2009, August). Active contour based lung field segmentation. In *2009 International Conference on Intelligent Human-Machine Systems and Cybernetics* (Vol. 1, pp. 288–291). IEEE.

18. Karoui, I., Fablet, R., Boucher, J. M., & Augustin, J. M. (2007, October). Unsupervised region-based image segmentation using texture statistics and level-set methods. In *2007 IEEE International Symposium on Intelligent Signal Processing* (pp. 1–5). IEEE.

19. Lavanya, M., & Kannan, P. M. (2017). Lung lesion detection in CT scan images using the fuzzy local information cluster means (FLICM) automatic segmentation algorithm and back propagation network classification. *Asian Pacific Journal of Cancer Prevention: APJCP, 18*(12), 3395.

20. Xue, W., Brahm, G., Pandey, S., Leung, S., & Li, S. (2018). Full left ventricle quantification via deep multitask relationships learning. *Medical Image Analysis, 43*, 54–65.

21. Yang Dong. *A Research on Coronary arteries Segmentation Algorithm for CTA Images and the Analysis on Vascular Stenosis Degree.* Zhejiang University, 2015.

22. Kavousi, M., Desai, C. S., Ayers, C., Blumenthal, R. S., Budoff, M. J., Mahabadi, A. A., ... & Khera, A. (2016). Prevalence and prognostic implications of coronary artery calcification in low-risk women: a meta-analysis. *JAMA, 316*(20), 2126–2134.

23. Mast, M. E., Heijenbrok, M. W., van Kempen-Harteveld, M. L., Petoukhova, A. L., Scholten, A. N., Wolterbeek, R., ... & Struikmans, H. (2016). Less increase of CT-based calcium scores of the coronary arteries. *Strahlentherapie und Onkologie, 192*(10), 696–704.

24. Gernaat, S. A., Išgum, I., De Vos, B. D., Takx, R. A., Young-Afat, D. A., Rijnberg, N., ... & Van Den Bongard, D. H. (2016). Automatic coronary artery calcium scoring on radiotherapy planning CT scans of breast cancer patients: reproducibility and association with traditional cardiovascular risk factors. *PLoS One, 11*(12), e0167925.

4 Deep Learning
A Review on Supervised Architectures and their Applications to Decision Support Systems in the Medical Field

*Tajinder Pal Singh, Sheifali Gupta,
Meenu Garg, and Deepali Gupta*

CONTENTS

4.1 INTRODUCTION

Artificial intelligence (AI) is a latest technology that deals with many applications in the domain of science, business, and government. It develops theories, methods, and techniques that simulate human intelligence and work to extend it (Smith and Eckroth 2017). AI is always seeking to develop a system that can understand the essence of intelligence and will enable that system to mimic the human brain.

Robotics, natural language processing, expert systems, and decision support systems are the various application areas of AI. Machine learning (ML) and deep learning (DL) both are subfields of AI. Multiple reviews have investigated and implemented ML and DL methods in healthcare and demonstrated the important effects of their improving efficiency and protection in healthcare (Miotto et al., 2018; Liang et al. 2014). Several ML and DL algorithms are used most commonly in medical field to solve clinical problems and help experts in making the diagnosis process more realistic and more accurate. The DL and ML methods can work as basis of expertise for computer aided decision support system which can further be an effective clinical decision support system (CDSS). It is a challenging task, however, but its development can provide help when making decisions and this could be part of the regular clinical workflow (Kawamoto et al. 2005). Therefore, CDSS has been recommended to facilitate decision-making, prevent medical mistakes, and improve patient health.

4.1.1 REVIEW METHODOLOGY

This chapter is split up into two sections; the first section discusses how several deep supervised architectures work, starting from regular neural network (NN), convolutional neural network (CNN), and through a recurrent neural network (RNN). This cluster in this chapter acts as a reference for additional readings or as a refresher for the preceding section that draws on this.

The second section of this chapter surveys the development and evaluation of DNN-based decision making support systems for detection, diagnosis, and classification of diseases in multiple medical fields (see Table 4.2). These medical fields relate to various diseases due to neurosurgerical, diabetes and heart problems in humans. For this chapter, publications are selected from Scopus and Pub-Med libraries. The range of 38 reviewed articles includes 13 articles belonging to diabetes mellitus, 10 articles on heart disease, and 15 articles on neurosurgery. For the articles selected in this review, the majority of researchers used accuracy, precision, sensitivity, and area under the curve (AUC) as performance indicators to evaluate the DNN-based decision support system.

4.2 DEEP SUPERVISED ARCHITECTURES

Classification using label data is known as supervised learning. During the data training of ML models, the output is generated for each group in the form of a score vector. There is little chance to score more in the category we want to classify than the other categories. This happenings during the initial stage of training. Therefore, an objective function is used to compute the error between the desired score and the output score. This objective function using internal parameters to rectify the error. This internal parameter is known as a weight. In comparison to ML models, DL models or deep supervised architectures contain a number of neural layers and millions of adjustable weights to modify the output score close to the desired score. Deep supervised architectures also work on large databases as compared to MLmodels because they don't performing well on small databases (Buczak and Guven 2016). Various deep supervised architectures are: artificial neural network, CNN, and RNN.

4.2.1 ARTIFICIAL NEURAL NETWORK

An artificial neural network (ANN) is basically parallel distributed processing, which can store the experimental knowledge for future use. The network acquires this knowledge through data learning processes and stores the same using the synaptic weight between the interneurons in different layers of the network.

The ANN is composed of three different sections; input layers, hidden layers, and output layers. The neurons of a layer are connected to each neuron of other layers through weights (W_{iI}, W_{iJ}, and W_{in} are representing the weights in different input layers of the network). In the learning process, each node receives the weighted output of all other nodes. They are first summed in the node and then activation function (α) is applied. The activated output is then passed to the next node's input. The same process is performed at each node of the whole NN during the feed forward mode.

In feed forwarding of the network, initially random weights are assigned to the NN, which mostly gives the output unlike the targeted output. The discrepancy generated between the expected output and the current NN output is called loss or error. This loss through error function is used to determine the criteria by which the weights in the network will be updated. The process of updating the weights in the network is known as back propagation (Rumelhart, Hinton, and Williams, 1985). It works for reducing the error through modifying the weights to improve the overall performance of the network. The error function used for reducing the loss is expressed by equation (4.1).

$$E = \frac{1}{2}\left[\Sigma_p \, \Sigma_i \left| t_{ip} - o_{ip} \right|^2 \right]^{1/2} \tag{4.1}$$

where E is root mean square (rms) value of error, t is targeted output and o is expected output over all pattern p. If the value of E is zero, all output patterns computed by the ANN is matched to desired value. If not, the modification of synaptic weights is done until the output come across the output node matched with the desired output. This whole modification is done using suitable learning methods. The weights after training contain some more meaningful information than the weights before training. This information is used in the future for data classification.

4.2.2 CONVOLUTIONAL NEURAL NETWORK

In DL, the convolutional neural network (CNN) is a multi-layers trained model connected end to end (Zeiler et al. 2010). A pipeline diagram of CNN layers is represented in Figure 4.1. The CNN is similar to normal NN with only difference; the neurons in the hidden layers of CNN are connected only to the subset of neurons in the previous layer. Due to this sparse connectivity, CNNs learn features implicitly.

Generally, CNNs have three main layers: a convolutional layer, a pooling layer, and a fully connected layer. The fully connected layer in actuality is an ANN that come after the other two main layers. The CNN can have many layers for each main layer. In other words, CNNs can be seen with one or many convolution, pooling, or fully connected layers. Each main layer of a CNN has its own function. Firstly,

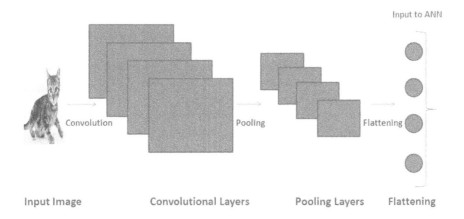

Convolution   Pooling   Flattening

Input Image   Convolutional Layers   Pooling Layers   Flattening

FIGURE 4.1   General architecture of a CNN.

the convolutional layer in a CNN is used to transform the input image and extract features from it. The input image is convolved with a kernel (or filter) in this image transformation. A kernel is a small matrix, smaller in height and width than the converted image. It is also called the matrix of convolution or the mask of convolution. Another construction block of a CNN is a pooling layer. It works for reducing the spatial size of the representation progressively to reduce network parameters and computations. The pooling layer performs independently for each function map. Max pooling is the most common method used in pooling. The third main layer is known as a fully connected layer or a general NN. Fully connected layers have many hidden layers with the last hidden layer known as loss layer. The loss layer computes the error between targeted and actual output value.

The CNN is trained in two different stages; forward stage and backward stage. In the forward stage of CNN training, the image is represented by the current parameter (weight and bias) in each layer. The value of the current parameter is initialized randomly during forward propagation. These values are treated as parameters from the CNN algorithm. After each forward propagation cycle, the loss is computed through the discrepancy between predicted output and true output at the output layer or the loss layer of the CNN. Different loss functions are used for calculating the loss or error. On the basis of the lost function output, backward propagation is performed. In the backward propagation process, internal parameters (weights) of the network are updated for computation of the next forward stage. The internal parameters in network training are updated in a manner to reduce the loss. After this kind of sufficient iteration of forward and backward propagation, the network training is stopped. The trained CNN has loss scaled to its minimum possible value. The next section will discuss the detail functioning of all main layers of CNN.

4.2.2.1 Convolutional Layer and Pooling Layer

A typical CNN architecture is organized in a sequence of stages. The starting stages are constituted by convolutional and pooling layers. These two layers form the basic unit of a CNN, involving most of the computation. Out of these two layers, the convolutional layer, is contained as sequence of various feature maps with neurons arranged inside. Filters or kernels are the learnable parameter of this layer. Each feature map of the convolutional layer contains rectified linear unit (ReLU)-activated values of the dot product of weights and local regions of the input to which neurons of the layer are connected. An input is an array that holds the image's raw pixel values. Here on input array or image, the filter or kernel is convoluted to produce output volume. The mathematical expression for convolution is given in equation (4.2), where the input array represents by f, kernel or filter by hand indexes of rows, and column in the resulting matrix by m and n.

$$G[m,n] = (f * h)[m,n] = \sum_j \sum_k h[j,k] f[m-j, n-k] \qquad (4.2)$$

Neurons that fall within the same feature map share the same weights, and keeping a limited number of parameters helps to reduce network complexity. On the other hand, distinct feature maps within a layer use different weights. Within an image, the local group of values is often highly correlated and the distinctive pattern they form can be easily detected. These patterns are also invariant to location in an image. Hence, whatever pattern of a correlated local group of values follows one part of an image, the same will be followed in the other part of an image. Thus, the idea of sharing the same weights by neurons at different locations of an image and detected the same pattern is highly appreciated. The benefits of the use of convolution are;

1. The various parameters are minimized by the feature maps using the weight-sharing method.
2. Local connectivity learns correlation with adjacent pixels.
3. Invariance to object locality

In an input array of an image, it's unrealistic to connect all neurons of a layer to input volume. Therefore, each neuron is only linked to the local area of an input. The spatial extent of this connectivity is a hyperparameter called a neuron's receptive region. It is equivalent to filter size. The hyperparameter that controls the output volume size and gives information about the neurons and their arrangement in the layer's output volume are depth, stride, and zero padding. Depth corresponds to the filter count chosen in the convolutional layer, where each filter in the convolutional layer learns something different from the input. Secondly, stride is all about filter movements. If stride is equal to 1, the filter moves over one pixel at a time, and if stride is equal to 2, filters move over two pixels at a time. In zero padding, all boundary pixels of input volume are padded with zeros. Zero padding helps to control the output volume's spatial scale. The formula used to calculate the size of output volume with respect to the size of input volume, receptive field size, strides, and zero padding is given in equation (4.3).

$$\frac{(W - F + 2P)}{S + 1} \qquad (4.3)$$

In equation (4.3), the W represents the input volume size, F represents the receptive field size, S represents stride, and P represents zero padding.

Although, the function of the convolutional layer is to detect local conjunctions of features from the previous layer, the pooling layer works for merging the similar features into one. Hence, pooling reduces the spatial dimension of representation, which leads to reduced network parameters and computational complexity. It also tends to tackle the over-fitting problem in the network. Various common pooling operations are max pooling, stochastic pooling (Zeiler and Fergus 2013), average pooling, spectral pooling (Rippel, Snoek, and Adams 2015), spatial pyramid pooling (Nguyen, Yosinski, and Clune 2015), and multiscale pooling (Gong et al. 2014). Out of these, max pooling is most commonly used. Mathematically, if a pooling layer accepts the input volume size W_{ip}, H_{ip}, D_{ip} (where W_{ip} is input width before pooling, H_{ip} is input height before pooling, and D_{ip} is input depth before pooling), then output volume size after pooling is W_{op}, H_{op}, D_{op} (where W_{op} is output width after pooling, H_{op} is output height after pooling, and D_{ip} is output depth after pooling) given in equation 4.4,4.5 and 4.6

$$W_{op} = \frac{\left(W_{ip} - F\right)}{S + 1} \tag{4.4}$$

$$H_{op} = \frac{\left(H_{ip} - F\right)}{S + 1} \tag{4.5}$$

$$D_{op} = D_{ip} \left(\text{remain Unchanged}\right) \tag{4.6}$$

The feature hierarchy constructed by two interleaved main layers (convolutional and pooling) of a CNN is transferred to several fully connected layers for final output of the network. This fully connected layer is part of the classification layer, followed by softmax function.

A fully connected layer of CNN is an ANN. The last layer of the fully connected network is known as loss layer. The loss layer computes the error occurring due to the difference between desired and actual value. The softmax function is commonly connected to this loss layer. The softmax function works based on probability distributions and is commonly used for predicting a single class out of n classes that are mutually exclusive. Some of the other loss functions mainly used in state vector machine (SVMs) and other large margin classifiers are hinge functions and squared hinge loss functions, etc.

4.2.3 RECURRENT NEURAL NETWORK

In CNNs, all inputs and outputs are considered to be independent from each other. But this will not work everywhere. In some cases, having connected information prevents the network from giving the desired results by treating that information independently. In such cases, a recurrent neural network (RNN) has been used often. RNNs work on sequential inputs or data. While accepting this sequential input, the RNN processes one element at a time and retains a state vector in its hidden unit that

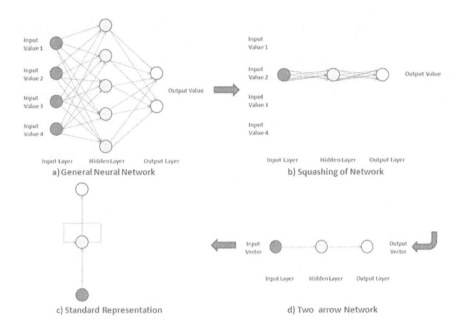

FIGURE 4.2 (a) Regular neural network, (b) Squashing of regular neural network, (c) Two-arrow network, and (d) Standard representation of an RNN.

implicitly contains historical knowledge of the past sequence element. Thus, the pre-dicted output of the RNN is based on current input and previous information. RNNs have a memory unit for holding the information predicated previously.

From an architectural point of view, an RNN is simply a general NN with a feed-back loop that enables the model to carry forward outputs from a previous neuron. In other words, artificial neurons in RNNs get their input from other neurons at previous steps, which make RNNs different from the general NN and make it possible to use it for complex tasks. Figure 4.2 represents the transformation of the regular NN to RNN.

Figure 4.2(a) represents a regular NN. If a regular NN is seen from underneath, it seems that the network is squashed as represented in Figure 4.2(b). The number of connections between neurons and the sequence of hidden layers does not change in the squashed network. The squashed network's multiple arrows are replaced with two arrows in Figure 4.2(c). After that, the vertical twisted two-arrow network with a temporal loop form an RNN as shown in Figure 4.2(d). This is the standard repre-sentation of an RNN, which shows that the hidden layer (temporal loop) in RNN not only gives an output but also feedback itself.

In RNN, the word recurrent implies that the network performed the same job on each element in a sequence based on output dependent on a previous computation. For that, RNN applies the recurrence rule to the present an input vector and its pre-vious state. Both of these are considered to have a present state. The formula for the present state is given in equation (4.7).

$$S_t = f(S_{t-1}, X_t) \tag{4.7}$$

where S_t is the present state, S_{t-1} represent previous state and X_t is presenting input. The equation simply depicts a standard representation of an RNN. In RNNs, each successive input coming into the network is considered to be a time step because the input neurons are applying a transformation on previous input. The simplest form of RNN in its present state at time t is represented in equation (4.8). Where tanh is an activation function; W_{ss} is the weight at the recurrent neuron; and W_{xs} is the weight at the input neuron.

$$S_t = tanh(W_{ss}S_{t-1}, W_{xs}X_t) \tag{4.8}$$

In the present case, the recurrent neuron only takes the immediate previous state. The same equation can be unfolded into multiple states for longer sequences. After calculating current state, the output state can be expressed by equation (4.9)

$$y_t = W_{sy}S_t \tag{4.9}$$

The output is compared with actual output and then the error is formed, which is then propagated back to the network to change the weights and train the network. This process is called back propagation. Back propagation has very exciting use in training the RNN. When hidden layer outputs in RNN are regarded at distinct discrete time steps as they are the outputs of DNN from different neurons, it becomes clear how exciting back propagation is for training RNN.

A RNN is a very powerful system, but its training is problematic due to exploding and vanishing gradients, during back propagation of time series data. Its rising due to shrinking and growing of back propagated gradients over each time step. Over several time steps, it will explode or vanish typically. In shorter contexts, RNNs will work efficiently, but when the context is longer, the information gets lost due to exploding or vanishing, which in turn makes RNNs less feasible. This problem can be solved easily with the help of a variant of RNNs called long short-term memory (LSTM) networks.

4.2.3.1 Long Short-Term Memory

The primary purpose of an RNN is to store long-term dependencies, but retaining knowledge over long periods of time is difficult (Bengio, Simard, and Frasconi 1994). LSTM networks with RNN as a hidden unit act naturally to recall long-term inputs. This hidden unit acts like a memory cell or accumulator, which is connected to itself at the next time step that has a weight of one; it copies its own real valued state and accumulates the external signal. A typical LSTM is shown in Figure 4.3.

In LSTM, C_{t-1} represents the memory cell input (at t time), X_t is an input (at t time), and h_t is an output (at t time) that goes both to output layer and the hidden layer in the proceeding time point. Hence, each block of LSTM has three inputs (X_t, h_{t-1}, and C_{t-1}) and two outputs (h_t and C_t) as shown in Figure 4.3. It should be noticed that all these inputs and outputs do not have single values, but vectors with lots of values behind each of them.

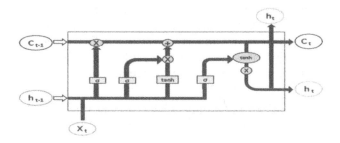

FIGURE 4.3   Long short-term memory unit.

LSTM is composed of various values: forget valve, memory valve, and output valve. Through these valves (X), a pipeline for the flow can be opened, closed, or partially opened. The forget valve will be closed, open, or partially open based on the decision of the sigmoid activation function (σ). If open, memory flows freely from C_{t-1} to C_t. If closed, memory is cut off and new memory will likely be added further in the pipeline. Sign "+"is a t-shaped joint, through which memory going through and additional memory can be added if the memory valve below this joint is open. The activation function "tanh" is responsible for transforming the value to be within the range from −1 to 1.

During LSTM operation, the X_t and h_{t-1} are received by the network as a current value and value from the previous node. Both values are combined and go through the sigmoid function. It will be determined at this stage whether the forgotten valve will be open, closed or partially open. The same values in parallel go through another two layers, the "tanh" and sigmoid layers operation. At "tanh" it will be determined which value will pass to the memory pipeline and at sigmoid layer operation it will be determined if that value is going to be passed to the memory pipeline and to what extent. Memory is flowing at the top of the pipeline. Here, if the forget valve is open with the memory valve closed, there is no change in memory. Otherwise, the memory will completely updated. So X_t and h_{t-1} together decide that what part of the memory pipeline will become the module's output.

Recently, DNNs have become a tool of choice in many applications or various disciplines. In health care, NNs are also used by many medical professionals for accurate analysis of disease, which helps them to treat that disease with better medical decisions. The next section represents an overview of different articles that applied NNs in various medical fields for decision support systems.

4.3   DECISION SUPPORT SYSTEMS

4.3.1   What Is Decision Support System?

A decision support system (DSS) is a computerized program that converts data into high quality information and can be used as a tool that supports decision-making activities. In general, a DSS is a computer-based information system for helping make decisions. A medical or clinical DSS (CDSS) is developed to help physicians enhance their efficiency at diagnosing various diseases related to neurosurgery, heart, diabetes, and hepatitis conditions, etc. (Kumar, Sathyadevi, and Sivanesh 2011). These systems are often based on static data. A CDSS system, which is developed to be learn the relationship between patient history, population illness, symptoms, disease diagnosis, family history, and test results, etc., is very useful to doctors and medical professionals. A CDSS can be composed of the various subsystems based on different applications; however the task is difficult. An identification of inadequate information for particular CDSS is a major challenge.

4.3.2 WHY ARE DECISION SUPPORT SYSTEMS SO IMPORTANT IN MEDICAL APPLICATIONS?

In the medical industry, quality treatment is desirable at an affordable cost. Quality treatment means the effective treatment of patients through correct diagnosis. A poor clinical decision may lead to devastating and unacceptable results. It is understandable that a medical diagnosis is subjective. First, it depends upon the doctor's prediction based on his diagnosis. Second, and most common, is the data analyzed for a good prediction is usually enormous and sometimes unmanageable. In this context, in order to improve diagnostic time and accuracy, it has been desirable to build a reliable and efficient CDSS to support the current increasingly difficult diagnosis-decision process.

4.3.3 SURVEY OF RECENT PUBLICATIONS

DSS is characterized as a framework based on prior information designed to help decision-makers use the appropriate data model to identify problems, solve them, and decide further action. The DSS is known to improve effectiveness, but not decision efficiency. Until now, many studies have been published focusing on medical diagnosis. The present state of publications include a survey on the development and evaluation of DSS for detection, diagnosis, and classification of neurosurgerical, diabetes, and heart disease issues. Diabetes is chronic disease and a major health issue worldwide. The International Diabetes Federation reported that about 246 million people suffer from diabetes and 3.8 million die every year from its complications (Kumar, Sathyadevi, and Sivanesh 2011). Heart disease is also a leading cause of death worldwide. Coronary artery disease (CAD) is a term for the narrowing or blockage in the coronary arteries of the heart. Many heart patients with CAD show symptoms like chest pain and fatigue, but some patients can show no symptoms. In 2012, 31% of total deaths worldwide was due to heart disease, which totaled 17.5 million deaths. Within the number of these deaths, 7.4 million peoples died from CAD alone. CAD is a major cause of death in Western countries (Mozaffarian et al., 2015.). Neurosurgery is also one of the highly risky fields that aim to reduce risks and improve surgical operations and outcomes. Currently, neurosurgeons have access to several data sources before, during, and after surgery. With this data, ML and DL have the ability to be used in better ways to improve patient neurosurgical health and increase the probability of successful outcomes. Keeping these three highly risky

TABLE 4.1
Reviewed Article Distribution According to Their Applications

Application areas	Authors
DSS for diabetes mellitus detection and diagnosis	(Mohebbi et al. 2017); (Kanungo, Srinivasan, and Choudhary 2017); (Lekha and Manikandan 2018); (Kwasigroch, Jarzembinski, and Grochowski 2018); (Benzamin and Chakraborty 2018.); (Arora and Pandey 2019); (Arcadu et al. 2019); (Karthikeyan et al. 2019); (Voets, Møllersen, and Bongo 2019); (Naz and Ahuja 2020); (Gadekallu et al. 2020); (Hemanth et al., 2020); (Zhou, Myrzashova, and Zheng 2020).
DSS for heart disease	(Hongmei Yan et al. 2003); (Haraldsson, Edenbrandt, and Ohlsson 2004); (Yan et al. 2006); (Tägil et al. 2008); (Lomsky et al. 2008); (Das, Turkoglu, and Sengur 2009); (Uğuz 2012); (Miao and Miao 2018); (Kose et al., 2021); (Ali et al., 2020).
DSS for neurosurgery	(Shi et al. 2013);(Mitchell et al. 2013);(Oermann et al. 2013);(Skrobala and Malicki 2014);(Azimi et al. 2014); (Azimi et al. 2015);(Baumgarten et al. 2016); (Dumont 2016); (Halicek et al., 2017); (Halicek et al. 2018); (Kim et al. 2018); (Kim et al., 2018); (Staartjes et al. 2019; (Hale et al. 2019; (Panesar et al. 2019).

fields in mind, the goal of the present study is to give an overview on the development and evaluation of deep supervised technique-based DSS for detection, diagnosis, and classification of diabetes mellitus, heart diseases, and neurosurgerical issues and to support the increasingly difficult diagnosis-decision process. The distribution of reviewed articles according to their application areas are presented in Table 4.1.

4.3.4 DEEP NEURAL NETWORK-BASED DECISION SUPPORT SYSTEM

All ML and DL algorithms have their own benefits and limitations. But NNs tend to work well as predictive tools possibly because they can model dynamic nonlinear interactions reliably in high-volume datasets. Previously, few review articles were published on automated decision-making system for diabetes mellitus, heart disease, and neurosurgerical condition detection and diagnosis based on various ML algorithms (Gupta and Chhikara 2018; Choudhury and Gupta 2019; Chaki et al. 2020; Safdar et al. 2018; Marimuthu et al. 2018; Chandra Reddy et al. 2019; Celtikci 2017; Senders et al. 2018; Buchlak et al. 2019). These research papers are published either on diabetes mellitus detection and diagnosis, heart disease detection and classification, or neurosurgery. In this attempt, a survey of the development and evolution of DSS for all these medical fields (diabetes mellitus, heart disease, and neurosurgery) is presented. The selected review articles are relevant as they provide a comprehensive overview of recent work in a specific field. The articles are taken from popular databases, i.e., Scopus and PubMed. Table 4.2 represents, DNN-based DSS for diabetes mellitus, heart disease, and neurosurgerical disease detection and diagnosis. The first two columns in Table 4.2 represent the DSS application areas and selected deep supervised approach. In few of studies, researchers used only one approach

TABLE 4.2
Research Articles that Applied DNNs to Facilitate Decision Making in Diabetes Mellitus, Heart Disease, and Neurosurgery

Application Area	Approach	Article Reference	Purpose/Field of Use	Dataset	Comparison	Results
		(Mohebbi et al. 2017)	Detection of type 2 diabetics	Continuous glucose monitoring signals from 9 different patients	Linear regression (LR) and multilayer perceptron (MLP)	Accuracy: 77.5%
		(Kanungo, Srinivasan, and Choudhary 2017)	Detection of diabetic retinopathy	Multiple datasets for training and Kaggle DR for testing	n/a	Accuracy: 88%
		(Lekha and Manikandan 2018)	Detection of diabetes mellitus	Dataset collected from MOS, the sensory unit	State vector machine (SVM) and neural network	Accuracy: 96.5%
		(Kwasigroch et al., 2018)	Detection of stages of diabetic retinopathy	EyePACS, LLC dataset	n/a	Accuracy: 81.7%
		(Benzamin and Chakraborty 2018)	Identification of hard exudates in fundus images	IDRiD dataset	n/a	Accuracy: 96.6%
DSS for diabetes mellitus	CNN	(Arora and Pandey 2019)	Detection and classification of diabetic retinopathy	Kaggle DR	n/a	Accuracy: 74.4%

Reference	Aim	Dataset	Methods	Results
(Arcadu et al. 2019)	Measurement of diabetic macular thickening	Kaggle DR dataset	n/a	Accuracy: 97%
(Karthikeyan et al. 2019)	Detection of multi-class retinal diseases	Diaretdb0 database, stare dataset and PIDD	n/a	Accuracy: 92%
(Voets, Møllersen, and Bongo 2019)	Classification of fundus image	EyePACS, LLC	n/a	Accuracy: 95.1%
(Naz and Ahuja 2020)	Prediction for risk measurement of diabetes	PIMA dataset	Artificial neural network (ANN), naive bayes (NB), Decision tree (DT)	Accuracy: 98.07%
(Gadekallu et al. 2020)	Detection of diabetic retinopathy	UCI machine learning repository	SVM, DT, k- nearest neighbor (KNN), NB, XG boost	Accuracy: 96%
(Hemanth, Deperlioglu, and Kose 2020)	Detection of diabetic retinopathy	MESSIDOR database	SVM, KNN, NB, Ensemble and other models	Accuracy: 97%, Sensitivity: 94%, Specificity: 98%,
(Zhou, Myrzashova, and Zheng 2020)	Diabetes Prediction	Self-prepared dataset and PIMA dataset	n/a	Accuracy: 99.4%

(Continued)

TABLE 4.2 (*Continued*)
Research Articles that Applied DNNs to Facilitate Decision Making in Diabetes Mellitus, Heart Disease, and Neurosurgery

Application Area	Approach	Article Reference	Purpose/Field of Use	Dataset	Comparison	Results
	Multilayer perceptron	(Hongmei Yan et al. 2003)	Heart disease diagnosis	352 cases of heart disease (86 for hypertension, 82 coronary heart disease, 71 rheumatic valvular heart disease, 60 chronic cor pulmonale, 53 congenital heart disease)	n/a	Accuracy: 82.9%
	Bayesian ANN	(Haraldsson, Edenbrandt, and Ohlsson 2004)	Detection of acute myocardial infarction (AMI)	2238 ECGs reports	ST-T amplitudes and slopes	Accuracy: 84.3%
	MLP (multilayer perceptron-)	(Yan et al. 2006)	Heart disease diagnosis	heart diseases database of 352 cases	n/a	Accuracy: >90%
DSS for heart diseases	Artificial neural network (ANN)	(Tägil et al. 2008)	Description of MPS	97 MPS studies of patients	Nuclear medicine specialists	Accuracy: 86%
	Neural Network	(Lomsky et al. 2008)	Interpretation of myocardial perfusion scintigraphy MPS	Training dataset of 418 MPS from one hospital and testing dataset of 532 MPS from another hospital	Software package ECTb	Sensitivity: 90% Specificity:85%

(Continued)

Method	Reference	Application	Dataset	Algorithm	Results
Neural Network Ensemble	(Das, Turkoglu, and Sengur 2009)	Diagnosis of valvular heart diseases	Dataset of 73 abnormal and 50 normal patients	FCM-CHMM	Sensitivity: 97.3% Specificity: 100%
ANN	(Uğuz 2012)	Classification sound signals from heart	Heart sounds from six patients using Littman 4100-model stethoscope	DFT-ANN	Accuracy: 95%
Deep Neural Network	(Miao and Miao 2018)	Diagnosis of coronary heart disease	Clinical Dataset of 303 patients	n/a	Accuracy: 83.67%,
Automatic encoder network	(Kose et al., 2021)	Early diagnosis of Heart Diseases	UCI learning dataset	ANN, SVM, Bayesian regulation algorithm	Accuracy: 99.13%, Sensitivity: 97.90% Specificity: 97.95%.
Ensemble deep learning	(Ali et al. 2020)	Monitoring system for heart disease	Two datasets of 303 and 294 cases	SVM, LR, MLP, RF, DT, NB	Accuracy: 98.5%
Neural network and LR	(Shi et al. 2013)	In-hospital mortality after traumatic brain injury (TBI) surgery	Dataset of 16,956 patients	LR	Accuracy: 95.61%
Neural network and LR	(Oermann et al. 2013)	1-year survival in patients with brain metastases treated with radiosurgery	Dataset of 196 patients	LR	Accuracy: 84%

TABLE 4.2 (*Continued*)
Research Articles that Applied DNNs to Facilitate Decision Making in Diabetes Mellitus, Heart Disease, and Neurosurgery

Application Area	Approach	Article Reference	Purpose/Field of Use	Dataset	Comparison	Results
DSS for neurosurgery	Neural networks	(Skrobala and Malicki 2014)	Beam orientation in stereotactic radiosurgery	Data of 539 patients	ANN networks	ANN2 produced suitable beam orientations
	Neural Network or neural network and LR	(Azimi et al. 2014)	2-year surgical satisfaction	Dataset of 168 patients	LR	Accuracy: 96.9%
	Neural network and LR	(Azimi et al. 2015)	Predicion of Recurrent lumbar disk herniation	Dataset of 402 patients	LR	Accuracy: 94.1%
	Neural network with combination of other classification techniques	(Baumgarten et al. 2016)	Occurrence of pyramidal tract side effects	Dataset of 10 patients	n/a	0.78 index value between the prediction PyMAN and the labeled data.
	Neural networks	(Dumont 2016)	The occurrence of symptomatic cerebral vasospasm	Dataset of 25 patients	n/a	Sensitivity -100%, Specificity -84 %
	CNN	(Halicek et al. 2017)	Classification of cancerous thyroid and aerodigestive tract tissues	Dataset of 50 head and neck cancer patients	SVM, KNN, LR, DTC, LDA	Accuracy: 96.4%, Sensitivity:96.8% Specificity: 96.1%
	CNN	(Halicek et al. 2018)	Distinguishing between thyroid carcinoma and normal tissue	Dataset of 21 head and neck cancer patients	n/a	Accuracy: 81% Sensitivity: 81% Specificity: 80%

Method	Reference	Task	Dataset	Comparison	Results
Neural network and LR	et al. 2018)	VTE, cardiac and wound complications, mortality	Dataset of 4,073 patients	LR	ANN outperformed LR in predicting cardiac complication, wound complication, and mortality
Neural network and LR	(Kim, et al., 2018)	VTE, cardiac and wound complications, mortality	Dataset of 22,629 patients	LR	ANN for accurate prediction of cardiac complications, and LR for accurate prediction of wound complications
Neural network and LR	(Staartjes et al., 2019)	Leg pain, back pain, and ODI	Dataset of 422 patients	LR	Accuracy: 85% for leg pain Accuracy: 87% for back pain Accuracy: 75% for functional disability
Neural network	(Hale et al., 2019	Predication of traumatic brain injury	Dataset of 12,902 patients	n/a	Accuracy: 97.98%
NN, LR, SVM, DT	(Panesar et al. 2019)	2-year mortality in glioma patients	Dataset of 76 patients	LR, SVM, DT	Accuracy: 70.7%

and did not compare the proposed classification technique to any other classification method. In the remained or the studies, where comparisons between classification methods were made, DNNs outperformed the other classification techniques. Specific tests were used for classification purposes in several studies and the results of the obtained values were seen with correctness measures improve results.

4.4   DISCUSSION

The work present in this article explored deep supervised techniques used in the development and evaluation of DSS for automated detection, diagnosis, and classification of neurosurgery, diabetes, and heart disease. In this review, most of publications are selected from Scopus and Pub-Med libraries. The range of 38 reviewed articles includes 13 articles belonging to diabetes mellitus, 10 articles on heart disease, and 15 articles belongs to neurosurgery.

The work done on diabetes mellitus can be categorized based on types of diabetes detected or classified by the technique applied. Mohebbi et al. (2017) and Naz and Ahuja (2020) worked on detection of type-II diabetics. Mohebbi et al. (2017) detected type 2 diabetics through adherence detection based on simulated continuous glucose monitoring (CGM) signals and Naz and Ahuja (2020) used DL on PIMA dataset for early-stage diabetes detection. Kwasigroch et al. (2018), Arora and Pandey (2019), Kanungo et al. (2017), and Hemanth et al. (2020) all worked on detection and classification of diabetic retinopathy based on feature extraction on color fundus photographs. Gadekallu et al. (2020) detected diabetic retinopathy using PCA (personal component analysis) for feature extraction and DL for classification. Lekha and Manikandan (2018) detected and classified diabetes mellitus based on real-time breath signals obtained from an array of gas sensors. Arcadu et al. (2019), Voets, Møllersen, and Bongo (2019), Benzamin and Chakraborty (2018), and Karthikeyan et al. (2019) worked on retinal fundus images for various purposes like measuring of diabetic macular thickening, fundus image classification, recognition of hard exudates in retina fundus images, and classification of fundus images. All of these articles feature similar shapes, color, texture, and automatically generated features used for diabetes mellitus research. Most of the articles used all three features, shape, color, texture (Kwasigroch et al. 2018; Arora and Pandey 2019; Arcadu et al. 2019; Kanungo, Srinivasan, and Choudhary 2017; Karthikeyan et al. 2019) and some of the articles used only two or a single type of feature in their work (Benzamin and Chakraborty 2018; Voets, Møllersen, and Bongo 2019; Lekha and Manikandan 2018). Datasets used in these articles are either self-created or freely accessible. Kaggle, EyePACS, LLC, and IDRiD datasets are the most commonly used freely accessible datasets. Accuracy, sensitivity, precision, and AUC are the various parameters on which researchers tried to justify their classifier. Out of these, all articles related to diabetes mellitus worked mostly on the accuracy of the model. The comparative analysis done in the various research articles proves that deep supervised architecture or CNN outperformed the other classification techniques.

In an overview of the DSS of heart diseases, performance of NN is compared and evaluates with gold standards and ML techniques. The majority of published work has identified the use of NN in DSS, which are used for understanding MPS in medical practice. The researchers identify ECG reports as having acute myocardial infarction (AMI), offer an overview of AMI factors in a graphical way, describe ischemia, and detect early AMI. In this study, one main limitation is that experiments are performed with limited data collected from various hospitals and labs. So to generalize the concept, more structured datasets with higher numbers of separate records for the patient is required for more accuracy.

The NN is the ideal option to facilitate decision-making before, during, and after neurosurgery. Found from the present study, NNs tended to outperform other algorithms across neurochirurgical professions, estimating correctly a number of operating outcomes when compared. The continued delivery of DL across well-built decision-support system complication levels are expected to be further reduced and neurochirurgical efficiency and health increased.

4.5   CONCLUSION

This chapter provides the detail overview of DNN-based DSS for automatic detection, diagnosis, and classification of neurosurgery, diabetes mellitus, and heart disease. CNN is mainly used for the automated recovery and analysis of diabetes mellitus results and ANN for neurochirurgical and heart diseases-based research problems. The majority of researchers used accuracy, sensitivity, precision, and AUC as metrics in terms of performance evaluation. In all these parameters, DNN-based DSS outperformed other techniques or methods. It can be concluded that continued delivery of DL across well-built decision-support system complication levels are expected to be further reduced and diagnostic process will be more accurate and effective.

REFERENCES

Ali, Farman, Shaker El-Sappagh, S.M. Riazul Islam, Daehan Kwak, Amjad Ali, Muhammad Imran, and Kyung-Sup Kwak. 2020. "A Smart Healthcare Monitoring System for Heart Disease Prediction Based on Ensemble Deep Learning and Feature Fusion," *Information Fusion* 63 (November): 208–22. https://doi.org/10.1016/j.inffus.2020.06.008.

Arcadu, Filippo, Fethallah Benmansour, Andreas Maunz, John Michon, Zdenka Haskova, Dana McClintock, Anthony P. Adamis, Jeffrey R. Willis, and Marco Prunotto. 2019. "Deep Learning Predicts OCT Measures of Diabetic Macular Thickening From Color Fundus Photographs." *Investigative Opthalmology & Visual Science* 60 (4): 852. https://doi.org/10.1167/iovs.18-25634.

Arora, Mamta, and Mrinal Pandey. 2019. "Deep Neural Network for Diabetic Retinopathy Detection." In *2019 International Conference on Machine Learning, Big Data, Cloud and Parallel Computing (COMITCon)*, 189–93. Faridabad, India: IEEE. https://doi.org/10.1109/COMITCon.2019.8862217.

Azimi, Parisa, Edward C. Benzel, Sohrab Shahzadi, Shirzad Azhari, and Hasan Reza Mohammadi. 2014. "Use of Artificial Neural Networks to Predict Surgical Satisfaction in Patients with Lumbar Spinal Canal Stenosis: Clinical Article." *Journal of Neurosurgery: Spine* 20 (3): 300–305. https://doi.org/10.3171/2013.12.SPINE13674.

Azimi, Parisa, Hassan R. Mohammadi, Edward C. Benzel, Sohrab Shahzadi, and Shirzad Azhari. 2015. "Use of Artificial Neural Networks to Predict Recurrent Lumbar Disk

Herniation:" *Journal of Spinal Disorders and Techniques* 28 (3): E161–65. https://doi.org/10.1097/BSD.0000000000000200.

Baumgarten, Clement, Yulong Zhao, Paul Sauleau, Cecile Malrain, Pierre Jannin, and Claire Haegelen. 2016. "Image-Guided Preoperative Prediction of Pyramidal Tract Side Effect in Deep Brain Stimulation: Proof of Concept and Application to the Pyramidal Tract Side Effect Induced by Pallidal Stimulation." *Journal of Medical Imaging* 3 (2): 025001. https://doi.org/10.1117/1.JMI.3.2.025001.

Bengio, Y., P. Simard, and P. Frasconi. 1994. "Learning Long-Term Dependencies with Gradient Descent Is Difficult." *IEEE Transactions on Neural Networks* 5 (2): 157–66. https://doi.org/10.1109/72.279181.

Benzamin, Avula, and Chandan Chakraborty. 2018. "Detection of Hard Exudates in Retinal Fundus Images Using Deep Learning," 2018 IEEE International Conference on System, Computation, Automation and Networking, Pondicherry, India, 2018, pp. 1–5). doi: 10.1109/ICSCAN.2018.8541246. 5.

Buchlak, Quinlan D., Nazanin Esmaili, Jean-Christophe Leveque, Farrokh Farrokhi, Christine Bennett, Massimo Piccardi, and Rajiv K. Sethi. 2019. "Machine Learning Applications to Clinical Decision Support in Neurosurgery: An Artificial Intelligence Augmented Systematic Review." *Neurosurgical Review*, 43 (August 2019): 1235–1253 https://doi.org/10.1007/s10143-019-01163-8.

Buczak, Anna L., and Erhan Guven. 2016. "A Survey of Data Mining and Machine Learning Methods for Cyber Security Intrusion Detection." *IEEE Communications Surveys & Tutorials* 18 (2): 1153–76. https://doi.org/10.1109/COMST.2015.2494502.

Celtikci, Emrah. 2017. "A Systematic Review on Machine Learning in Neurosurgery: The Future of Decision Making in Patient Care." *Turkish Neurosurgery*. https://doi.org/10.5137/1019-5149.JTN.20059-17.1.

Chaki, Jyotismita, S. Thillai Ganesh, S.K Cidham, and S. Ananda Theertan. 2020. "Machine Learning and Artificial Intelligence Based Diabetes Mellitus Detection and Self-Management: A Systematic Review." *Journal of King Saud University – Computer and Information Sciences*, July, S1319157820304134. https://doi.org/10.1016/j.jksuci.2020.06.013.

Chandra Reddy, N. Satish, Song Shue Nee, Lim Zhi Min, and Chew Xin Ying. 2019. "Classification and Feature Selection Approaches by Machine Learning Techniques: Heart Disease Prediction." *International Journal of Innovative Computing* 9 (1). https://doi.org/10.11113/ijic.v9n1.210.

Choudhury, Ambika, and Deepak Gupta. 2019. "A Survey on Medical Diagnosis of Diabetes Using Machine Learning Techniques." In *Recent Developments in Machine Learning and Data Analytics*, edited by Jugal Kalita, Valentina Emilia Balas, Samarjeet Borah, and Ratika Pradhan, 740:67–78. Advances in Intelligent Systems and Computing. Singapore: Springer Singapore. https://doi.org/10.1007/978-981-13-1280-9_6.

Das, Resul, Ibrahim Turkoglu, and Abdulkadir Sengur. 2009. *Computer Methods and Programs in Biomedicine* 93 (2): 185–91. https://doi.org/10.1016/j.cmpb.2008.09.005.

Dumont, Travis M. 2016. "Prospective Assessment of a Symptomatic Cerebral Vasospasm Predictive Neural Network Model." *World Neurosurgery* 94 (October): 126–30. https://doi.org/10.1016/j.wneu.2016.06.110.

Gadekallu, Thippa Reddy, Neelu Khare, Sweta Bhattacharya, Saurabh Singh, Praveen Kumar Reddy Maddikunta, In-Ho Ra, and Mamoun Alazab. 2020. "Early Detection of Diabetic Retinopathy Using PCA-Firefly Based Deep Learning Model." *Electronics* 9 (2): 274. https://doi.org/10.3390/electronics9020274.

Gong, Yunchao, Liwei Wang, Ruiqi Guo, and Svetlana Lazebnik. 2014. "Multi-Scale Orderless Pooling of Deep Convolutional Activation Features." In *Computer Vision – ECCV 2014*, edited by David Fleet, Tomas Pajdla, Bernt Schiele, and Tinne Tuytelaars,

8695:392–407. Lecture Notes in Computer Science. Cham: Springer International Publishing. https://doi.org/10.1007/978-3-319-10584-0_26.

Gupta, Ankita, and Rita Chhikara. 2018. "Diabetic Retinopathy: Present and Past." *Procedia Computer Science* 132: 1432–40. https://doi.org/10.1016/j.procs.2018.05.074.

Hale, Andrew T., David P. Stonko, Jaims Lim, Oscar D. Guillamondegui, Chevis N. Shannon, and Mayur B. Patel. 2019. "Using an Artificial Neural Network to Predict Traumatic Brain Injury." *Journal of Neurosurgery: Pediatrics* 23 (2): 219–26. https://doi.org/10.3171/2018.8.PEDS18370.

Halicek, Martin, Baowei Fei, James V. Little, Xu Wang, Mihir Patel, Christopher C. Griffith, Amy Y. Chen, and Mark W. El-Deiry. 2018. "Optical Biopsy of Head and Neck Cancer Using Hyperspectral Imaging and Convolutional Neural Networks." In *Optical Imaging, Therapeutics, and Advanced Technology in Head and Neck Surgery and Otolaryngology 2018*, edited by Brian J. F. Wong, Justus F. Ilgner, and Max J. Witjes, 33. San Francisco, United States: SPIE. https://doi.org/10.1117/12.2289023.

Halicek, Martin, Guolan Lu, James V. Little, Xu Wang, Mihir Patel, Christopher C. Griffith, Mark W. El-Deiry, Amy Y. Chen, and Baowei Fei. 2017. "Deep Convolutional Neural Networks for Classifying Head and Neck Cancer Using Hyperspectral Imaging." *Journal of Biomedical Optics* 22 (6): 060503. https://doi.org/10.1117/1.JBO.22.6.060503.

Haraldsson, Henrik, Lars Edenbrandt, and Mattias Ohlsson. 2004. "Detecting Acute Myocardial Infarction in the 12-Lead ECG Using Hermite Expansions and Neural Networks." *Artificial Intelligence in Medicine* 32 (2): 127–36. https://doi.org/10.1016/j.artmed.2004.01.003.

Hemanth, D. Jude, Omer Deperlioglu, and Utku Kose. 2020. "An Enhanced Diabetic Retinopathy Detection and Classification Approach Using Deep Convolutional Neural Network." *Neural Computing and Applications* 32 (3): 707–21. https://doi.org/10.1007/s00521-018-03974-0.

Hongmei Yan, Jun Zheng, Yingtao Jiang, Chenglin Peng, and Qinghui Li. 2003. "Development of a Decision Support System for Heart Disease Diagnosis Using Multilayer Perceptron." In *Proceedings of the 2003 International Symposium on Circuits and Systems, 2003. ISCAS '03.*, 5:V-709-V–712. Bangkok, Thailand: IEEE. https://doi.org/10.1109/ISCAS.2003.1206411.

Kanungo, Yashal Shakti, Bhargav Srinivasan, and Savita Choudhary. 2017. "Detecting Diabetic Retinopathy Using Deep Learning." In *2017 2nd IEEE International Conference on Recent Trends in Electronics, Information & Communication Technology (RTEICT)*, 801–4. Bangalore: IEEE. https://doi.org/10.1109/RTEICT.2017.8256708.

Karthikeyan, S., Sanjay Kumar P., R J Madhusudan, S K Sundaramoorthy, and P K Krishnan Namboori. 2019. "Detection of Multi-Class Retinal Diseases Using Artificial Intelligence: An Expeditious Learning Using Deep CNN with Minimal Data." *Biomedical & Pharmacology Journal* 12 (3): 1577–86. https://doi.org/10.13005/bpj/1788.

Kawamoto, Kensaku, Caitlin A Houlihan, E Andrew Balas, and David F Lobach. 2005. "Improving Clinical Practice Using Clinical Decision Support Systems: A Systematic Review of Trials to Identify Features Critical to Success." *BMJ* 330 (7494): 765. https://doi.org/10.1136/bmj.38398.500764.8F.

Kim, Jun S., Varun Arvind, Eric K. Oermann, Deepak Kaji, Will Ranson, Chierika Ukogu, Awais K. Hussain, John Caridi, and Samuel K. Cho. 2018. "Predicting Surgical Complications in Patients Undergoing Elective Adult Spinal Deformity Procedures Using Machine Learning." *Spine Deformity* 6 (6): 762–70. https://doi.org/10.1016/j.jspd.2018.03.003.

Kose, Utku, Omer Deperlioglu, Jafar Alzubi, and Bogdan Patrut. 2021. "A Practical Method for Early Diagnosis of Heart Diseases via Deep Neural Network." In *Deep Learning for Medical Decision Support Systems*, by Utku Kose, Omer Deperlioglu, Jafar Alzubi, and Bogdan Patrut, 909:95–106. Studies in Computational Intelligence. Singapore: Springer Singapore. https://doi.org/10.1007/978-981-15-6325-6_6.

Kumar, D Senthil, G Sathyadevi, and S Sivanesh. 2011. "Decision Support System for Medical Diagnosis Using Data Mining" 8 (3): 7.

Kwasigroch, Arkadiusz, Bartlomiej Jarzembinski, and Michal Grochowski. 2018. "Deep CNN Based Decision Support System for Detection and Assessing the Stage of Diabetic Retinopathy." In *2018 International Interdisciplinary PhD Workshop (IIPhDW)*, 111–16. Swinoujście: IEEE. https://doi.org/10.1109/IIPHDW.2018.8388337.

Lekha, S., and Suchetha M. 2018. "Real-Time Non-Invasive Detection and Classification of Diabetes Using Modified Convolution Neural Network." *IEEE Journal of Biomedical and Health Informatics* 22 (5): 1630–36. https://doi.org/10.1109/JBHI.2017.2757510.

Liang, Znaonui, Gang Zhang, Jimmy Xiangji Huang, and Qmming Vivian Hu. 2014. "Deep Learning for Healthcare Decision Making with EMRs." In *2014 IEEE International Conference on Bioinformatics and Biomedicine (BIBM)*, 556–59. Belfast, United Kingdom: IEEE. https://doi.org/10.1109/BIBM.2014.6999219.

Lomsky, Milan, Peter Gjertsson, Lena Johansson, Jens Richter, Mattias Ohlsson, Deborah Tout, Andries van Aswegen, S. Richard Underwood, and Lars Edenbrandt. 2008. "Evaluation of a Decision Support System for Interpretation of Myocardial Perfusion Gated SPECT." *European Journal of Nuclear Medicine and Molecular Imaging* 35 (8): 1523–29. https://doi.org/10.1007/s00259-008-0746-9.

Marimuthu, M., Abinaya, M., Hariesh, K. S., Madhankumar K. V. 2018. "A Review on Heart Disease Prediction Using Machine Learning and Data Analytics Approach." International Journal of Computer Applications 181 (18): 20–25. https://doi.org/10.5120/ijca2018917863.

Miao, Kathleen H, and Julia H. 2018. "Coronary Heart Disease Diagnosis Using Deep Neural Networks." *International Journal of Advanced Computer Science and Applications* 9 (10). https://doi.org/10.14569/IJACSA.2018.091001.

Miotto, Riccardo, Fei Wang, Shuang Wang, Xiaoqian Jiang, and Joel T Dudley. 2018. "Deep Learning for Healthcare: Review, Opportunities and Challenges." *Briefings in Bioinformatics* 19 (6): 1236–46. https://doi.org/10.1093/bib/bbx044.

Mitchell, Timothy J., Carl D. Hacker, Jonathan D. Breshears, Nick P. Szrama, Mohit Sharma, David T. Bundy, Mrinal Pahwa, et al. 2013. "A Novel Data-Driven Approach to Preoperative Mapping of Functional Cortex Using Resting-State Functional Magnetic Resonance Imaging." *Neurosurgery* 73 (6): 969–83. https://doi.org/10.1227/NEU.0000000000000141.

Mohebbi, Ali, Tinna B. Aradottir, Alexander R. Johansen, Henrik Bengtsson, Marco Fraccaro, and Morten Morup. 2017. "A Deep Learning Approach to Adherence Detection for Type 2 Diabetics." In *2017 39th Annual International Conference of the IEEE Engineering in Medicine and Biology Society (EMBC)*, 2896–99. Seogwipo: IEEE. https://doi.org/10.1109/EMBC.2017.8037462.

Mozaffarian, Dariush, Emelia J Benjamin, Alan S Go, Donna K Arnett, Michael J Blaha, Mary Cushman, Sarah de Ferranti, et al. (2015). "Heart Disease and Stroke Statistics—2015 Update," Circulation 131 (4): 297.

Naz, Huma, and Sachin Ahuja. 2020. "Deep Learning Approach for Diabetes Prediction Using PIMA Indian Dataset." *Journal of Diabetes & Metabolic Disorders* 19 (1): 391–403. https://doi.org/10.1007/s40200-020-00520-5.

Nguyen, Anh, Jason Yosinski, and Jeff Clune. 2015. "Deep Neural Networks Are Easily Fooled: High Confidence Predictions for Unrecognizable Images." In *2015 IEEE Conference on Computer Vision and Pattern Recognition (CVPR)*, 427–36. Boston, MA, USA: IEEE. https://doi.org/10.1109/CVPR.2015.7298640.

Oermann, Eric K., Marie-Adele S. Kress, Brian T. Collins, Sean P. Collins, David Morris, Stanley C. Ahalt, and Matthew G. Ewend. 2013. "Predicting Survival in Patients With

Brain Metastases Treated With Radiosurgery Using Artificial Neural Networks." *Neurosurgery* 72 (6): 944–52. https://doi.org/10.1227/NEU.0b013e31828ea04b.

Panesar, Sandip S., Rhett N. D'Souza, Fang-Cheng Yeh, and Juan C. Fernandez-Miranda. 2019. "Machine Learning Versus Logistic Regression Methods for 2-Year Mortality Prognostication in a Small, Heterogeneous Glioma Database." *World Neurosurgery: X* 2 (April): 100012. https://doi.org/10.1016/j.wnsx.2019.100012.

Rippel, Oren, Jasper Snoek, and Ryan P Adams. 2015. "Spectral Representations for Convolutional Neural Networks," (29th Conference on Neural Information Processing Systems, Montreal, Canada, 2015).

Rumelhart, D E, G E Hinton, and R J Williams. 1985. "Learning Internal Representations by Error Propagation," (Technical Report 8506). Institute for Cognitive Science, University of California, San Diego.

Safdar, Saima, Saad Zafar, Nadeem Zafar, and Naurin Farooq Khan. 2018. "Machine Learning Based Decision Support Systems (DSS) for Heart Disease Diagnosis: A Review." *Artificial Intelligence Review* 50 (4): 597–623. https://doi.org/10.1007/s10462-017-9552-8.

Senders, Joeky T., Patrick C. Staples, Aditya V. Karhade, Mark M. Zaki, William B. Gormley, Marike L.D. Broekman, Timothy R. Smith, and Omar Arnaout. 2018. "Machine Learning and Neurosurgical Outcome Prediction: A Systematic Review." *World Neurosurgery* 109 (January): 476–486.e1. https://doi.org/10.1016/j.wneu.2017.09.149.

Shi, Hon-Yi, Shiuh-Lin Hwang, King-Teh Lee, and Chih-Lung Lin. 2013. "In-Hospital Mortality after Traumatic Brain Injury Surgery: A Nationwide Population-Based Comparison of Mortality Predictors Used in Artificial Neural Network and Logistic Regression Models: Clinical Article." *Journal of Neurosurgery* 118 (4): 746–52. https://doi.org/10.3171/2013.1.JNS121130.

Skrobala, Agnieszka, and Julian Malicki. 2014. "Beam Orientation in Stereotactic Radiosurgery Using an Artificial Neural Network." *Radiotherapy and Oncology* 111 (2): 296–300. https://doi.org/10.1016/j.radonc.2014.03.010.

Smith, Reid G., and Joshua Eckroth. 2017. "Building AI Applications: Yesterday, Today, and Tomorrow." *AI Magazine* 38 (1): 6. https://doi.org/10.1609/aimag.v38i1.2709.

Staartjes, Victor E., Marlies P. de Wispelaere, William Peter Vandertop, and Marc L. Schröder. 2019. "Deep Learning-Based Preoperative Predictive Analytics for Patient-Reported Outcomes Following Lumbar Discectomy: Feasibility of Center-Specific Modeling." *The Spine Journal* 19 (5): 853–61. https://doi.org/10.1016/j.spinee.2018.11.009.

Tägil, K., M. Bondouy, J. P. Chaborel, W. Djaballah, P. R. Franken, S. Grandpierre, B. Hesse, et al. 2008. "A Decision Support System Improves the Interpretation of Myocardial Perfusion Imaging." *European Journal of Nuclear Medicine and Molecular Imaging* 35 (9): 1602–7. https://doi.org/10.1007/s00259-008-0807-0.

Uğuz, Harun. 2012. "A Biomedical System Based on Artificial Neural Network and Principal Component Analysis for Diagnosis of the Heart Valve Diseases." *Journal of Medical Systems* 36 (1): 61–72. https://doi.org/10.1007/s10916-010-9446-7.

Voets, Mike, Kajsa Møllersen, and Lars Ailo Bongo. 2019. "Reproduction Study Using Public Data of: Development and Validation of a Deep Learning Algorithm for Detection of Diabetic Retinopathy in Retinal Fundus Photographs." Edited by Sandra Ortega-Martorell. *PLOS ONE* 14 (6): e0217541. https://doi.org/10.1371/journal.pone.0217541.

Yan, H, Y Jiang, J Zheng, C Peng, and Q Li. 2006. "A Multilayer Perceptron-Based Medical Decision Support System for Heart Disease Diagnosis." *Expert Systems with Applications* 30 (2): 272–81. https://doi.org/10.1016/j.eswa.2005.07.022.

Zeiler, Matthew D., and Rob Fergus. 2013. "Stochastic Pooling for Regularization of Deep Convolutional Neural Networks." *ArXiv:1301.3557 [Cs, Stat]*, January. http://arxiv.org/abs/1301.3557.

Zeiler, Matthew D., Dilip Krishnan, Graham W. Taylor, and Rob Fergus. 2010. "Deconvolutional Networks." In *2010 IEEE Computer Society Conference on Computer Vision and Pattern Recognition*, 2528–35. San Francisco, CA, USA: IEEE. https:// doi.org/10.1109/CVPR.2010.5539957.

Zhou, Huaping, Raushan Myrzashova, and Rui Zheng. 2020. "Diabetes Prediction Model Based on an Enhanced Deep Neural Network." *EURASIP Journal on Wireless Communications and Networking* 2020 (1): 148. https://doi.org/10.1186/s13638-020-01765-7.

5 Automated View Orientation Classification for X-ray Images Using Deep Neural Networks

K. Karthik and Sowmya Kamath S

CONTENTS

5.1 INTRODUCTION

The availability of advanced diagnostic technologies in medical imaging has resulted in large volumes of medical image data. As the capacity of storing these images grows, searching for relevant images when required for clinical and diagnostic task-related decision making has become an uphill task. Medical image management systems that aid in effective retrieval of clinically relevant information or knowledge from medical images, while also facilitating the use of such knowledge toward supporting intelligent applications like clinical decision support systems (CDSS), predictive analytics systems, diagnostic applications, etc. are a critical need for leveraging this large-scale diagnostic data in an impactful way [1, 2]. Radiological procedures like X-rays have evolved, which is a crucial diagnostic imaging tool for identifying abnormalities in different body parts, which may require insights derived from various views/body orientations of the patient. Often, a *frontal view* and *lateral view* are used in such cases. For computer-aided diagnosis (CAD), internal and external shapes are very important in identifying the abnormality. Currently, the projection view/image orientation of radiographs are labeled manually by radiologists and technicians. Manual corrections for wrongly labeled views make it impractical in Picture Archive and Communication System (PACS) and digital imaging systems, because it involves cost and time of human resources. Instead of manually labeling such multi-oriented images, it can be accomplished automatically by intelligent algorithms that are trained to understand the patterns with large-scale images [3]. Methods that can assess this automatically and provide the necessary information regarding the view of the organ at which the scan is taken can be beneficial. Therefore, a classifier model developed for categorizing the disease according to the image view is of great importance. Further, this helps in providing a proper description of the image in an overall clinical workflow management system.

While scanning, i.e., during the diagnostic image capturing process, the scanning equipment is focused on the injured part of the body, and scans are typically performed in different positions to aid effective diagnosis. Hence, there is a need for developing an effective model on view classification based on the image orientation. In solving the challenges for view classification, most of the approaches were developed by using traditional hand-crafted features, such as local binary patterns (LBP), scale-invariant feature transform (SIFT) features, histogram of oriented gradients (HOG). There are very limited works when it comes to neural network-based orientation classification. Ahn et al. [4] used CNN and DNN models on a head pose dataset consisting of 15,678 upper body images of 20 people for pose prediction. The network structure achieved good performance and computational speed compared to other existing algorithms. Pose classification [5] is another work that focused on determining the orientation of movement of the pedestrians. They reported that visual low-level features without attention mechanisms were able to achieve good performance.

The objective of this work is to design a model for view orientation classification with reference to the different orientations in which the patients' diagnostic scans are performed, during the initial process of diagnostic scanning. The outcome of this should be accurate, continuous, operating in the real time in the clinical workflow CAD systems. We achieve this by exploiting different deep neural network architectures and demonstrate that the proposed methodology outperforms with other experimental works carried here, and we benchmark our results. Our approach is adequate for real-time applications during preprocessing and facilitate the use of it in building intelligent applications like CDSS, predictive analytical systems, etc.

This chapter is presented as follows: Section 5.2 briefs on existing works related to orientation view recognition-based approaches for diagnostic scan images. Section 5.3 details the proposed methodology for orientation detection and classification, and the specifics of the neural network models designed for the task. Section 5.4 describes the experimental analysis and observations along with the results of the proposed CNN-based models. We also present directions for further improvements in the proposed system.

5.2   RELATED WORK

Identifying the body orientations from medical scan images is an essential requirement for accurately indexing and categorizing the incoming image data in large-scale hospital information management systems (HIMS) [6]. As per our observations, only a few works exist currently that focus on chest radiograph images. Scans of other body parts and modalities where tissue orientation is also a significant need have not been addressed adequately and this is a research gap that needs to be considered.

Hence, there is a scope for view classification for other biological structures that can be incorporated as an essential step during the indexing process applied to scanned radiograph images, thus aiding the overall management of HIMS. Existing works that address the challenge of view classification are discussed here.

Arimura et al. [7] proposed a set of nine templates, one set for medium-sized patients consisting of three templates (1 PA and 2 lateral) and another set for small/ large-sized patients consisting of six templates (2 PA and 4 lateral). The similarity of the chest image in lieu of one of these templates was determined using a template matching technique, considering a correlation value greater than 0.2. Their approach was a two-step process for identifying the orientations of the image; in the first step, two different views were determined for medium-sized patients. If it is unidentified, then a check with the other set involving six templates is performed in the second step. A total of 1,000 test images were used in their experiments, involving 500 PA and 500 lateral chest radiograph images. In the first step, 924 cases (92.4%) were correctly identified, while all other cases were identified in the second step. Lehmann et al. [8] determined chest radiographs' view by applying several distance measures and nearest-neighbor classification. Using tangent distance as the nearest-neighbor classification scheme, good accuracy was obtained for images of 32×32 pixels. Boone et al. [9] proposed a feed-forward neural network to identify views in chest X-ray images, in which a series of chest images consisting of 999 lateral and 999 frontal were down sampled to a size of 16×16 during training. The network was able to identify the views of chest images with 98.8% on an average of six trails. Kao et al. [10] developed a projection profile technique to identify frontal and lateral views for chest X-ray images. The projection profile was computed based on the computation of two indices, namely body symmetry index and background percentage index.

During chest X-ray screening, the orientation view information is a crucial aspect. Santosh and Wendling [11] developed a novel method for classifying the chest X-ray image view as frontal and lateral. They incorporated a new technique called angular relational signature to extract features from the histogram. Multilayer perceptron, random forest, and support vector machine were used to predict the classification accuracy, attaining close to 100%. Kao et al. [12] developed an automatic recognition of frontal PA and AP chest radiographs. Their work incorporated three features in identifying the chest radiographic views, i.e., the scapula's and clavicle's tilt angles and the extent of radiolucency in the lung. The method was evaluated with 1,200 chest radiographs, consisting of 600 PA and 600 AP images. The performance was measured with receiver operating characteristic (ROC), which illustrated that the fusion of the previously mentioned three features showed a high discriminant result. Takeuchi et al. [13] developed an automated chest X-ray radiography classification for the CheXpert dataset consisting of 65,240 patient images, labeled by an expert radiologist. The work explored different network architectures and found that the featured DenseNet121 passed into a decision tree classifier achieved an accuracy of 93%.

Scanned radiographs are stored frequently in PACS with an unknown orientation label, making it ineffective for radiologist analysis. A solution to this problem was proposed by Luo et al. [14] with an automated protocol for chest images. Desired regions of the chest features were extracted, such as its size, rotation, and translation, and a trained classifier was used for identifying the directional view of the chest images. The alignment label was then distinguished considering the abdomen and the neck positions in the radiograph image. The experiment showed promising results of about 96%, with 6,680 images collected from a hospital. Xue et al. [15] developed a hybrid feature model for chest X-rays categorizing the images into frontal and lateral. The experiment was performed on two datasets—the NLM Indiana and the IRMA datasets, consisting of 8,000 images. Combined features of image profile (IP), contour-based shape feature (CBSF), and pyramid of histograms of orientation gradients (PHOG) with a 10-fold cross-validation achieved a good accuracy when used with CAD systems. However, from Santosh and colleagues' work [16], it is observed that the algorithm was trained with frontal chest X-rays, where it won't classify the lateral chest images. Certain features are also essential when classifying both the views of an image. The primary reason is that the features (shape and texture) vary with both the frontal and lateral view images. Another novel technique developed by Santosh et al. [17] for identifying the rotated lungs in chest X-rays measures the rib orientation using a generalized line histogram technique for quality control.

From the review of the few existing works that attempt view classification, it was observed that there is significant scope for extensive research in this area. While archiving medical images in HIMS, the parts of the body that are scanned at varied angles require proper categorization of orientation labels. To achieve this, the supervised algorithms need to be trained on similar images of the same part of the body with correct orientation labels, which in itself is a challenging task. An automatic system that recognizes the orientation view label of different parts of the body soon after the scan is taken will add value to HIMS, aiding in efficient indexing, categorization, and storage. Designing such methods will be beneficial in reducing the incidence of mislabeled or unlabeled images in the overall medical image management process. In the next section, we discuss the specifics of the proposed approach for view classification.

5.3 PROPOSED METHODOLOGY

The sequence of tasks involved in the overall progress of the proposed view orientation classification model using a pretrained network is shown in Figure 5.1. We adopt neural- network architectures as an alternative to traditional supervised learning-based systems, due to their adaptive learning behavior and semi-supervised nature. Four different convolutional neural networks are used as transfer learning approaches for optimizing the training process. Apart from this, we also designed a new neural model called ViewNet for the task of medical image view classification. The objective is to use neural models that are capable of learning the ample feature representations for a large number of image views, by incorporating transfer learning approaches for identification of the orientation label for different body part images. Transfer learning techniques are generally used in most deep learning applications. The advantage of using a pretrained network is to learn a modern task, making it easier and quicker than learning the features from scratch. Another benefit here is that the model learns better with even lesser number of training images, when trained for newer tasks with the transfer learning approaches.

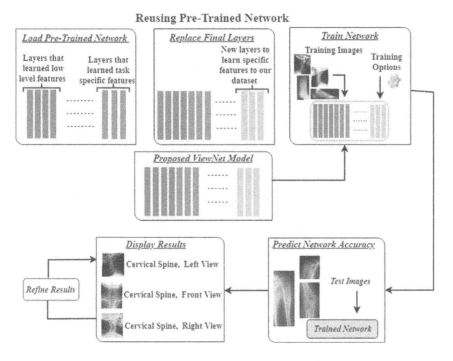

FIGURE 5.1   Proposed approach for view/body orientation classification.

For the orientation identification task, a subset of orientation-identified classes from the ImageCLEF 2009 dataset [18] are used. We used 41 different classes consisting of different organs of the body like the spine (cervical spine, thoracic spine, lumbar spine), leg, hand, carpal bones, nose, and eye area. Each class had its own IRMA code, which is considered for the orientation classification task (described in detail in Section 5.4). Images of size $m \times n$ are fed into the neural network for training. Each dataset image had different image dimensions, therefore resizing was performed as soon as the images were read from the datastore, i.e., before feeding it into the network. Augmentation was also performed on the training dataset through operations like random flip and translation along the vertical and horizontal axis, which prevents the network from overfitting. While training different deep neural network architectures using a transfer learning approach, some of the final layers should be changed to a fully connected, a softmax, and a classification output layers by observing the dataset used. The newly replaced fully connected layer parameters need to be specified according to the new dataset for a new classification model. Increasing the *Weight Learn Rate Factor* and *Bias Learn Rate Factor* values helps the network learn features faster, with the addition of the new layers. Also while training the model, several hyperparameter values like mini-batch, epochs, batch normalization, learning rate, regularizations, optimizers, and activators were varied, and finally the best suitable values were chosen in building the final model for the orientation identification task.

We used four different neural models for our initial benchmarking experiments—AlexNet, ResNet18, GoogleNet, and SqueezeNet. The specifics of these models are discussed here. AlexNet [19] comprises of 25 layers, with the first layer as the image input layer having 227×227 dimension, followed by the first convolution layer with a window shape of 11×11. Since most medical scan images have larger dimensions than natural photographic images, more pixels are required to copy the data present in the image. Consequently, a larger convolution window is used in the next layer for handling the input data. Later, the convolution window size is reduced to 5×5 and 3×3, respectively. Next to the convolutional layer, rectified linear unit (ReLU) activation and a max-pooling layer are placed excluding at Conv3 and Conv4. A max-pooling and a normalization layer reside in between the Conv1 and Conv2. Two fully connected layers are present after the last convolutional layer that produces feature maps with size 4096. Finally, a softmax followed by a classification layer is used for the prediction. Compared to other CNNs, the main difference is that AlexNet comprises of more convolution channels.

ResNet18 [20] is 71 layers deep with 224×224 dimension for image input, followed by the first convolution layer with a window shape of 7×7, and a max-pooling layer. A total of 20 such convolution layers are present in this network with different batches. The convolution window shape in the first batch layer is of 1×1, whereas in the second batch layer it is of 3×3. Between each convolution layer, batch normalization and ReLU activation function are placed, except at the first and second convolution layer, which has an additional max-pooling layer. Only at the last convolution layer pooling layer changes as an average pooling layer. The network architecture concludes by a fully connected, a softmax, and a classification output layer.

GoogleNet [21] is 144 layers deep with 224×224 dimension for image input. Convolution layers of Conv1 and Conv3 have ReLU activation function, max-pooling and cross-channel normalization layers in batches. A total of 57 such convolution layers are present in this network with different batches. Every convolution layer has a ReLU activation function, the max-pooling layer is added for a batch of each sixth convolutions starting from the eighth convolution layer. The next preceding convolution layer has a depth concatenation layer, with the same height and width of the convolution layer and concatenates them along the channel dimension. The convolution layer window size varies from 5×5, 1×1, 1×1, 1×1, 3×3, 1×1 in the batch of each six convolution layers. The last convolution block ends with an average pooling layer and a dropout by 40%. The network architecture ends with a fully connected, a softmax, and a classification output layer. SqueezeNet [22] is 68 layers deep with 227×227 dimension for image input. The first convolution layer has a window shape of 3×3, with a ReLU activation function along with a max-pooling layer. A total of 26 such convolution layers are present in this network with different batches. Starting from the Conv4 layer for every third convolution layer a depth concatenation layer is added. The convolution layer's window size varies in its number from 3×3, 1×1, 1×1. The last convolution block ends with a ReLU activation function and an average pooling layer. The architecture of the network completes by adding a fully connected, a softmax, and a classification output layer.

We also developed a new CNN architecture adapted from AlexNet and ResNet18, called ViewNet, since it is used for the classification of image-based orientation

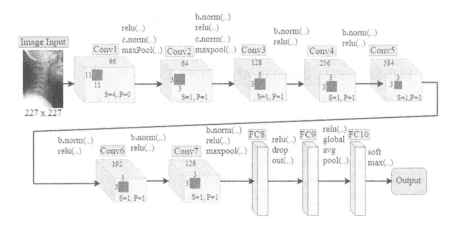

FIGURE 5.2 Proposed architecture of ViewNet.

views. The network comprises of 35 layers having an image input layer with a dimension of 227 × 227. A complete architectural model of the ViewNet is represented in Figure 5.2. The first convolution layer has a window shape of 7 × 7, followed by a ReLU activation function, cross-channel normalization, and a max-pooling layer. Further, the CNN is further built up with a grouped convolution layer, batch normalization, ReLU activation function, cross-channel normalization, and a max-pooling layer. Next, convolution, batch normalization, and the ReLU activation function are used five times, and finally a max-pooling layer is added prior to the first fully connected layer. In between the fully connected layers, dropout, the ReLU activation function, global average pooling are also added. The network architecture ends with fully connected, softmax, and a classification output layer.

Table 5.1 lists the hyper parameter values determined during training for each of the deep neural network models as discussed in Section 5.3. The optimization function used in all four networks used for the benchmarking experiments and the proposed ViewNet is stochastic gradient descent momentum (SGDM) algorithm, which works faster and better than stochastic gradient descent. The main advantage of SGDM is that it helps to accelerate the gradient's vectors in the right direction,

TABLE 5.1

Classification Model Parameters

NN Model	Learning Rate	Weight/Bias	Batch Size	Epochs
ViewNet (proposed)	0.0001	20	8	10
AlexNet [19]	0.0001	20	8	10
ResNet18 [20]	0.0001	10	8	8
GoogleNet [21]	0.0003	10	10	6
SqueezeNet [22]	0.0001	20	10	8

leading to faster convergence. We describe the experiments conducted to validate the proposed approach and observations regarding the performance in the next section.

5.4   EXPERIMENTAL RESULTS AND DISCUSSION

For the experimental validation, we used the ImageCLEF 2009 dataset [18], which consists of 41 unique classes of different orientations of body organs like spine (i.e., cervical spine, thoracic spine, umbar spine), leg, hand, carpal bones, nose, and eye area. Some classes had all three different orientations (e.g., spine) while some classes had only two orientations (e.g., nose and eye area). A set of sample images that are taken for this work from the dataset are shown in Figure 5.3.

The IRMA code consists of a 13-character unique code, which is subdivided into four parts along its axis. The notation appears as TDAB, where *T* refers to *technical* (4 digits); *D*, *directional* (3 digits); *A*, *anatomical* (3 digits); and *B*, *biological* (3 digits). Considering *D* (i.e., body orientation) and *A* (i.e., examined part of the body) subparts of the IRMA code, the directional/view orientation identification label is predicted for various body parts by using different neural network models. The three-digit directional code (DDD) gives a detailed description of the image orientation view. The first digit gives details about the orientation (e.g., *1-coronal, 2-sagittal, 3-axial*). The second digit gives more information about the position (e.g., *11-posteroanterior (PA), 12-anteroposterior (AP)*) and the third digit details the direction on the orientation of the type of the organ examined (e.g., *218-inclination*).

Similarly, the three-digit anatomical code (AAA) gives a detailed description of which part of the body was examined. Major regions are coded as (e.g., *1-whole body, 2-cranium, 3-spine, 4-upper extremity/arm, 5-chest, 6-breast, 7-abdomen, 8-pelvis, 9-lower extremity/leg*) following a two hierarchical subcodes (e.g., *7-abdomen, 71-upper abdomen, 711-upper right quadrant, 712-upper middle quadrant, 713-upper left quadrant*). Using these codes of observation helps in building a proper model for the orientation identification system.

| IRMA Code ➔ | 1121-110-213-700 | 1121-420-213-700 | 1121-220-213-700 | 1121-210-213-700 |
| Code Description ➔ | Nose Area - Front | Nose Area - Front | Nose Area - Left | Nose Area - Right |

| IRMA Code ➔ | 1121-220-310-700 | 1121-120-310-700 | 1121-218-310-700 | 1121-210-310-700 |
| Code Description ➔ | Cervical Spine - Left | Cervical Spine - Front | Cervical Spine - Right | Cervical Spine - Right |

FIGURE 5.3   Sample dataset images showing the IRMA code and its description.

The proposed CNN models' performance are measured using the standard metrics like accuracy, sensitivity, specificity and F1 score as per Eq. (5.1) to (5.4). Accuracy is computed as the total number of accurate predictions on view class labels divided by the total number of images under test classification. Precision is the number of correctly predicted orientation view labels from a set of selected images. In contrast, recall is the total number of correctly predicted orientation view labels from the entire test set. F1 score helps to have a measurement that represents both (i.e., TPR and TNR), and it is the weighted average of the true positive rate and true negative rate.

$$Accuracy = \frac{TP + TN}{TP + TN + FP + FN} \tag{5.1}$$

$$Precision = \frac{TP}{TP + FP} \tag{5.2}$$

$$Recall = \frac{TP}{TP + FN} \tag{5.3}$$

$$F1\ Score = \frac{2 * TP}{(2 * TP + FP + FN)} \tag{5.4}$$

The classification results before refining, i.e., the classes that have three different orientations/views (front view, lateral views—right and left) is depicted in Table 5.2. However, it was observed that some of the class IRMA codes are different, but the orientations/views do not vary. In such cases, we combined both the IRMA codes to a single label (i.e., we performed refining). After all such class labels are combined, we again classify the test images, after which a significant improvement in accuracy is observed. The classification results observed after the process of refining the classes was performed are shown in Table 5.3.

TABLE 5.2
Observed Classification Performance w.r.to Different CNN Models
(Before Class Label Refinement)

NN Model	Accuracy	Precision	Recall	F1 Score
ViewNet (proposed)	0.8571	0.5387	0.5841	0.5298
AlexNet [19]	0.8549	0.5636	0.5882	0.5634
ResNet18 [20]	0.8521	0.5033	0.5161	0.4987
GoogleNet [21]	0.6543	0.3788	0.3994	0.3609
SqueezeNet [22]	0.6502	0.3332	0.3420	0.3082

TABLE 5.3

Observed Classification Performance w.r.to Different CNN Models *(After Class Label Refinement)*

NN Model	Accuracy	Precision	Recall	F1 Score
ViewNet (proposed)	0.9151	0.5743	0.5996	0.5414
AlexNet [19]	0.9145	0.6361	0.7011	0.7411
ResNet18 [20]	0.9182	0.5731	0.5971	0.5741
GoogleNet [21]	0.9053	0.5491	0.6191	0.5531
SqueezeNet [22]	0.8868	0.5141	0.5921	0.5161

Experimental evaluation revealed that AlexNet and ResNet18 achieved good classification accuracy of 85.49% and 85.21% (before merging the same orientation classes) and 91.45% and 91.82% (after merging). Based on this result, we also developed our own CNN model named ViewNet, adapted from AlexNet and ResNet CNN models. The experimental results of this model outperformed all the other CNN models used in the benchmarking experiments. It was found that the proposed ViewNet reached an accuracy of 85.71% with a small improvement over AlexNet. This improvement was attained because the layers that are incorporated in the proposed ViewNet model are taken from both of the top resulted models of AlexNet and ResNet. More importantly, the complexity of the network is significantly lower than AlexNet and ResNet, while still achieving comparable results. The compact architecture actually helps achieve significant optimization in training time, which can be highly significant when the size of medical image repositories that are to be is large. We intend to tune the proposed ViewNet for further improvements in performance, so that it can accurately predict diagnostic images' body orientation for large-scale HIMS as well.

5.5   CONCLUSION AND FUTURE WORK

The proposed work deals with an efficient and accurate method for identifying the orientation label of the body organ positioned at the time the scan is developed. We used four types of transfer learning-based neural models for benchmarking the orientation classification task on the standard open dataset, ImageCLEF 2009. A novel architecture, ViewNet, was also proposed for the task of view classification. The five neural models were validated on the images available from the dataset, with its IRMA code specifics, and achieved promising results when measured in terms of accuracy, sensitivity, specificity, and F1 score. The possible applications of this work are in the context of HIMS systems, for effective and accurate view labeling the scan images after the scan process, to optimize the indexing process, thus streamlining the workflow of HIMS. As part of future work, we plan to extend the proposed system with reference to increasing the number of organs along with three different views, so that a more robust and hospital-friendly system can be incorporated, to further enhance the overall management of medical systems.

ACKNOWLEDGMENTS

The authors gratefully acknowledge the Science and Engineering Research Board, Department of Science and Technology, Government of India, for its financial support through the Early Career Research Grant (Grant no. ECR/2017/001056) and the facilities at the Department of Information Technology, NITK Surathkal.

REFERENCES

1. Vikram, M., Anantharaman, A., Suhas, B.S., Kamath, S.S., An approach for multimodal medical image retrieval using latent dirichlet allocation. In: Proceedings of the ACM India Joint International Conference on Data Science and Management of Data. (2019) 44–51.
2. Soundalgekar, P., Kulkarni, M., Nagaraju, D., Kamath, S.: Medical image retrieval using manifold ranking with relevance feedback. In: 2018 IEEE 12th International Conference on Semantic Computing (ICSC), IEEE (2018) 369–373
3. Karthik, K., Kamath, S.S.: A deep neural network model for content-based medical image retrieval with multi-view classification. The Visual Computer (2020) 1–14
4. Ahn, B., Park, J., Kweon, I.S.: Real-time head orientation from a monocular camera using deep neural network. In: Asian conference on computer vision, Springer (2014) 82–96
5. Gepperth, A., Ortiz, M.G., Heisele, B.: Real-time pedestrian detection and pose classification on a gpu. In: 16th International IEEE Conference on Intelligent Transportation Systems (ITSC 2013), IEEE (2013) 348–353
6. Karthik, K., Kamath, S.S.: A hybrid feature modeling approach for content- based medical image retrieval. In: 2018 IEEE 13th International Conference on Industrial and Information Systems (ICIIS), IEEE (2018) 7–12
7. Arimura, H., Katsuragawa, S., Li, Q., Ishida, T., Doi, K.: Development of a computerized method for identifying the posteroanterior and lateral views of chest radiographs by use of a template matching technique. Medical Physics 29(7) (2002) 1556–1561
8. Lehmann, T.M., Güld, O., Keysers, D., Schubert, H., Kohnen, M., Wein, B.B.: Determining the view of chest radiographs. Journal of Digital Imaging 16(3) (2003) 280–291
9. Boone, J.M., Hurlock, G.S., Seibert, J.A., Kennedy, R.L.: Automated recognition of lateral from pa chest radiographs: saving seconds in a pacs environment. Journal of Digital Imaging 16(4) (2003) 345–349
10. Kao, E.F., Lee, C., Jaw, T.S., Hsu, J.S., Liu, G.C.: Projection profile analysis for identifying different views of chest radiographs. Academic radiology 13(4) (2006) 518–525
11. Santosh, K., Wendling, L.: Angular relational signature-based chest radiograph image view classification. Medical & biological engineering & computing 56(8) (2018) 1447–1458
12. Kao, E.F., Lin, W.C., Hsu, J.S., Chou, M.C., Jaw, T.S., Liu, G.C.: A computerized method for automated identification of erect posteroanterior and supine anteroposterior chest radiographs. Physics in Medicine & Biology 56(24) (2011) 7737
13. Takeuchi, D., Thai, R., Tran, K.: Exploring model architectures and view-specific models for chest radiograph diagnoses. http://cs229.stanford.edu/proj2019spr/report/17.pdf (2019) 1–6
14. Luo, H., Hao, W., Foos, D.H., Cornelius, C.W.: Automatic image hanging protocol for chest radiographs in pacs. IEEE Transactions on Information Technology in Biomedicine 10(2) (2006) 302–311
15. Xue, Z., You, D., Candemir, S., Jaeger, S., Antani, S., Long, L.R., Thoma, G.R.: Chest x-ray image view classification. In: 2015 IEEE 28th International Symposium on Computer-Based Medical Systems, IEEE (2015) 66–71

16. Santosh, K., Vajda, S., Antani, S., Thoma, G.R.: Edge map analysis in chest x-rays for automatic pulmonary abnormality screening. International journal of computer assisted radiology and surgery 11(9) (2016) 1637–1646
17. Santosh, K., Candemir, S., Jaeger, S., Karargyris, A., Antani, S., Thoma, G.R., Folio, L.: Automatically detecting rotation in chest radiographs using principal rib-orientation measure for quality control. International Journal of Pattern Recognition and Artificial Intelligence 29(02) (2015) 1557001
18. Lehmann, T.M., Schubert, H., Keysers, D., Kohnen, M., Wein, B.B.: The irma code for unique classification of medical images. In: Medical Imaging 2003: PACS and Integrated Medical Information Systems: Design and Evaluation. Volume 5033., International Society for Optics and Photonics (2003) 440–451
19. Krizhevsky, A., Sutskever, I., Hinton, G.E.: Imagenet classification with deep convolutional neural networks. In: Advances in neural information processing systems. (2012) 1097–1105
20. He, K., Zhang, X., Ren, S., Sun, J.: Identity mappings in deep residual networks. In: European conference on computer vision, Springer (2016) 630–645
21. Szegedy, C., Liu, W., Jia, Y., Sermanet, P., Reed, S., Anguelov, D., Erhan, D., Vanhoucke, V., Rabinovich, A.: Going deeper with convolutions. In: Proceedings of the IEEE conference on computer vision and pattern recognition. (2015) 1–9
22. Iandola, F.N., Han, S., Moskewicz, M.W., Ashraf, K., Dally, W.J., Keutzer, K.: Squeezenet: Alexnet-level accuracy with 50x fewer parameters and 0.5 mb model size. arXiv preprint arXiv:1602.07360 (2016)

6 Sustainable e-Health Solutions for Routine and Emergency Treatment
Scope of Informatics and Telemedicine in India

Arindam Chakrabarty, Uday Sankar Das, and Saket Kushwaha

CONTENTS

6.1 INTRODUCTION

Achieving superior health resources is a reflection of the overall commitment and attitude of the state to its citizens. Health policy has to be designed for a holistic perspective not from a narrow dimension. Improved healthcare systems reflect and reveal the essence of economic development. A healthcare system does not mean good quality of treatment, creating innovative molecules, or advance diagnostic systems. It represents the well-crafted policy coupled with its effective and efficient implementation in society to ensure the optimized quality of life, enjoyed by the people in general with a minimized average morbidity period. The healthcare ecosystem has been greatly fueled by threshold technological advancement that enables quality health solutions with users. Despite achieving healthcare excellence baring a few, a large section of the society is unable to access and avail the benefits. It is

73

because of high-cost integration, lack of flexibility or convenience, and serious concerns regarding logistic issues that act as stumbling blocks for deriving appropriate and timely medical treatment.

On the contrary, the common people have no information, awareness, or even knowledge of how to and where to go to treat a specific health problem. Surprisingly there is no exchange of information or data sharing among the health service providers such as hospitals, diagnostic centers, etc. Whenever a patient visits a new hospital or new physician, the healthcare facility starts from scratch without having any prior information about the patient or previous treatment delivered. This results in an average standard of healthcare service that kills time, increases cost, and impacts badly on the economy as a whole.

It becomes a Herculean task, a gigantic responsibility, for any government to create a state-of-the-art infrastructure for providing healthcare service for every citizen, particularly for a country with an elephantine population. In the post-economic reform period, private players are also operating in providing health services, but their services are confined to the creamy layer of the society who can afford the high cost for their treatment. A country like India needs inclusiveness in healthcare services where a person at the lowest stratum would be provided quality services for treating health-related problems on par with people belonging to other economic segments.

The benefits of telemedicine can have a huge impact on society if a centralized database is created as a policy for the country, where a detail of all the previous and ongoing patient records are available [1]. Similarly, big data analytics paired with artificial intelligence (AI) have a tremendous potential to provide healthcare insights about an individual's health on a real-time basis while alerting the patient [2] to take necessary precautions before any health-related complications take a toll on life. E-health is a potential tool to bridge the gap between the availability of doctors per patient and free up the burden on hospital infrastructure in a country like India [3].

6.1.1 GLIMPSE OF E-SOLUTION

The inception of digital technology into the healthcare scenario amplified the possibilities through which healthcare services could be delivered while eliminating the hurdles posed by the geographic barriers and remoteness. A holistic e-health solutions environment may include services such as electronic health records (EHR), health providers directory (HPD), terminology server (TS), master patient index (MPI), and hospital information management system (HIMS).These digital healthcare infrastructures can be augmented to deliver various workflows to achieve health excellence [4]. The World Health Organization (WHO) classifies digital health interventions (DHIs) into the following sections: (1) clients (user-oriented information and patient services), (2) healthcare providers (health practitioner–oriented health services), (3) health system managers (healthcare manager–oriented services), and (4) data services (health system manager–oriented services) [5].

Some of the challenges in the implementation of health IT systems include government spending in tele-health, infrastructure, digital literacy, interoperability, etc. Despite these challenges, subsequent governments have been focused on bringing health IT to the public health network, for example, the Department of

Information Technology has invested its efforts in establishing general telemedicine, teleradiology, electronic portal imaging devices (EPIDs), multileafcollimators (MLCs), 6-Mev medical linear accelerators, etc. The Ministry of Health and Family Welfare, Government of India, has also implemented projects like OncoNET India, Integrated Disease Surveillance Project (IDSP), Teleophthalmology, National Rural Telemedicine Project, National Medical College Telemedicine Network, etc. under the umbrella of e-health initiatives [6].

These technology-led interventions help the citizens of the country to obtain both routine and emergency health services in India.

6.1.2 Overview of Health Informatics in India

Health IT (H-IT) adoption in India is lagging despite the abundance of the available workforce. Training and development still remains a bottleneck for this transformative miracle to be achieved in India in the future. Public health spending has traditionally been at the lower end, much lower than the WHO recommended percentage. H-IT services like telemedicine have the potential for a resultant savings of 80% of healthcare spending particularly in rural communities [7]. Health informatics roughly consist of components like hospital information systems (HISs), human resource management information systems (HRMISs), health management information systems(HMISs), geographic information systems (GISs), and mobile health applications and monitoring systems [8].

Experts in the Wharton India Economic Forum 2020, Mumbai, highlighted several positive aspects of India's overall healthcare initiatives powered by artificial intelligence and telemedicine, despite several lacunas that preexist. The three Bs of health care, known as biology, bytes, and bandwidth, are playing a crucial role in the growth of e-health in India, where technology-oriented biology is helping reduce the cost of testing and diagnostics, bytes and the ever-increasing computing power is helping shape technologies that will eventually reduce the cost to a minimum, and bandwidth or the telecommunications network is reducing the digital divide to access health care.

Efforts like Make in India has also played a vital role in reducing the cost of imported medical devices like stents and implants in a significant way. These technology-led interventions are helping bridge the gap of doctor-to-patient ratio in India, for instance, Ask-Apollo can reduce the patient wait time to 20 minutes only, and1mg provides AI-based consultation for about 20% of its users. It is evident that a holistic approach toward e-health will have positive growth for the health sector in India while making cost-effective and timely health interventions for citizens living in poverty, remote locations, or inaccessible terrains [9].

6.2 LITERATURE REVIEW

This literature review is based on contemporary articles subject to their availability and generic application of e-health and telemedicine as a potential health tool. Also, various dimensions of security and difficulty in the application of e-health are explored.

The COVID pandemic highlighted the importance for mechanisms that potentially can help mitigate the risks of being within the proximity of the infected patient. Information communication technology (ICT) tools and particularly telemedicine have the potential to reduce and minimize the dangers while the consolidation of such tools will have a huge impact on integrated healthcare [10].

Developing nations have tremendous difficulty in providing access to free health care to the majority of their populations. Telemedicine can play a role in bridging this gap and thereby increasing effective treatment. The Technology Acceptance Model has the potential to help improve the healthcare scenarios of the developing countries across the globe while helping the policymakers take the right direction in the adoption of telemedicine [11].

In the United States, at least 15% of physicians are equipped with telemedicine in their workplace. In the United Kingdom, the National Health Service has adopted a vision to take telemedicine mainstream. Telemedicine will see a potential boom when it is adopted by middle- and low-income groups to replace in-person care with virtual visits that further lower the cost and increase the convenience. Virtual visits are not just convenient these days but are safer due to the nature of the new normal. Countries such as the United States, Germany, and Norway already have telemedicine in the form of ambulance units equipped with telestroke services for stroke patients. Virtual visits can provide a holistic integration of doctors, nurses, therapists, and dietitians, all-round diagnostics and patient care [12].

The COVID pandemic has proven the need and reliance of even developed societies in telemedicine. There was an exceptional increase in telemedicine consultation for nonurgent healthcare as the threat of physical proximity or contact with any potential COVID patient could increase the risk of infection several folds [13].

In the new normal, palliative care as well as clinical medicine has gone through transformative changes with telemedicine at the core of providing healthcare service that reduces three-sided risks for patients, patient's families, and the clinician. Telemedicine reduces the burden of personal protective equipment any healthcare institution faces in pandemic scenarios. A systematic and step-by-step implementation of telemedicine can become a boon for any nation [14].

Digital healthcare services use cloud servers for storing data that is used by both the patients and healthcare providers. Cloud technology is also vulnerable to threats from malicious programs, hence cryptographic and non-cryptographic mechanisms may help reduce the vulnerability and bring about a holistic change in obtaining foolproof security and sustainability [15]. Similarly, a combination of private blockchain and consortium blockchain holds the key to promise security of personal health information (PHI) achieved through e-health blockchains [16].

Seamless digital healthcare applications depend heavily on wireless technologies, however, electromagnetic interference and the possibility of malfunction caused by radio frequency (RF) transmission remains a real threat in a mobile hospital scenario. A power control algorithm is proposed as a solution to overcome these scenarios [17].

Immunology can also be referred to as a data-rich discipline because it involves patient health data analysis and continuous interaction between healthcare professionals and patients. Mobile computing has invented the buzz word *m-health*, which covers a broad domain of public health practices useful for monitoring patient device information. This phenomenon has also raised a question of data accuracy in public health. The burden of innovation remains on the shoulders of the immunologists for the continuous development of potential apps to propagate digital health [18].

E-health is a potential tool to mitigate all sorts of losses that may be caused by a pandemic through its control and surveillance mechanisms. Switching over to e-health requires proper change management planning. A comprehensive e-health readiness framework may be the answer to reduce the impact of pandemic situations [19].

VIRTUS IoT middleware is an alternative to legacy service-oriented architecture. Thus it is an effective tool to monitor a patient's body movements and daily activities as a modular architecture deployed in a wide scenario of e-health solutions [20].

Ubiquitous technologies hold the promised healthcare revolution for chronic and critical diseases by providing telemedicine solutions. Flex-RFID middleware may help in providing flexible and convenient healthcare solutions [21].

The artificial immunity system through its self-learning and adaptation mechanisms can mitigate the information system threat [22].

Any novel influenza strain has the potential to cause a pandemic as was evident from the lessons of the last century. E-health applications for public health surveillance can help in outbreak investigation and designing appropriate responses [23].

Global interoperability regulations are needed to overcome the risk posed to patient health data. Health information organizations must lead the way in creating a professional and certified health data security protocol on a global level with an appropriate code of ethical conduct [24].

6.3   OBJECTIVES OF THE STUDY

1. To study the emergence and status of treatment through an online mode in the form of e-health strategy particularly in consonance with the lessons learned during the COVID-19 pandemic.
2. To explore the meaningful synergy between health informatics and telemedicine to improvise e-health strategy in India.
3. To recommend comprehensive decentralized appropriate policy reforms and implementation strategy in India for achieving health indicative UNSDGs.

6.4   RESEARCH METHODOLOGY

The research is based on secondary information. However, attempts are made to identify the emergence and significance of popularizing health informatics and a telemedicine network in India particularly in the recent experience of the pandemic disaster.

6.5   ANALYSIS AND INTERPRETATION

6.5.1   ANALYSIS I

NSS 71st Round depicts that the average nonmedical expenditure per event hospitalization during the last 365 days was Rs. 2021 and Rs. 2019 in rural and urban areas across the country compared to the average medical expenditure of Rs. 14,935 and

Rs. 24,436 in rural and urban areas respectively. This shows that nonmedical expenditures that includes transportation expenditure on the escort, food, lodging, and other charges like stationers and toiletries, etc. In case of prolonged hospitalization, the nonmedical expenditure escalates to a roaring high. The nonmedical expenditures are also incurred for nonhospitalization cases like routine consultation performing diagnostic tests, etc. The nonmedical expenditure is more for rural people becausethey have to commute to the city or urban area for availing superior medical facilities or support.There are numerous cases where it becomes difficult to escort a patient immediately due to nonfullfilling of feasibility criteria, for example, getting a train or flight reservation, not having instant financial resources, managing leave for the patient or escort on a short period of notice, etc. All these results in serious compromise on public health issues.

Indicative impediments of the present health ecosystem may be categorized as follows:-

- Increase of overall household expenditure on health as it is conjugated by increasing nonmedical expenditure(71st NSS Round)
- Compromise on timely treatment due to various feasibility constraints.
- Huge financial loss for the patient or patient party.
- Loss of man-days for the patient escort.
- It becomes tantamount if the treatment is continued for a prolonged period.
- Present health ecosystem considering the growing phenomenon for patient mobility issues for availing superior treatment(inclusive of in-house and outdoor treatment)

From this discussion and understanding the surface reality in the health domain, it is imperative to improvise a comprehensive e-health strategy in India so that the elephantine population spread across the country's urban, rural, and remote regions can be provided equitable health services either through strengthing the physical infrastructure or widening the ambit of virtual resources. If a set of the population from a disadvantaged or remote area cannot be provided an adequate treatment facility through physical mode or resources, it is high time to implement strategic e-health solutions to render appropriate medical treatment on a real-time basis. Otherwise, the average morbidity of people willincrease, which, in turn, retards the engine of economic growth. This is unaffordable for India particularly in the ongoing and upcoming economic disaster due to the severe outbreak of the COVID-19 pandemic [25].

Recent reports indicate that there is an improving pattern of access to telemedicine in India. According to the Statista website, 2018, the opinion of using telemedicine are as follows (Table 6.1).

This shows that there is huge potential for future telemedicine users because around 58% of the respondents had a positive outlook toward telemedicine.

The National Health Portal of India lists several significant efforts made by the government of India toward the digitation of health care. The following list depicts the same along with the details (Table 6.2).

Several private firms also provide e-health services in India as shown in Table 6.3.

TABLE 6.1

Telemedicine Usage in India 2018 [26]

Telemedicine Usage Opinion	
Used telemedicine services and will use it again	17%
Did not like telemedicine services and will not use again	10%
Would like to try the services	41%
Would not even give telemedicine a try	15%
Not sure about using telemedicine	17%

TABLE 6.2

Indicative List of Telemedicine Services Provided by the Government of India [27]

Name of the Service	Details of the Service and Applications
Disease A–Z	Health Terminologies Directory
Online Registration System (ORS)	Electronic Health Record System
mCessation	Rehabilitation Services For Tobacco Addicts
NHP-Voice Web (1800-180-1104)	24/7 Toll-Free Helpline
Mobile Applications	NHP Health Directory Services
	NHP Swasth Bharat
	NHP Indradhanush Immunization
	No More Tension
	Pradhan Mantri Surakshit Matritva Abhiyan (PMSMA)
	Mera Aspataal (My Hospital)
	India Fights Dengue
	National AIDS Control Organization App
	Jansankhya Sthirata Kosh (JSK)
	Journey of First 1,000 Days (Ayushman Bhava)
e-Rakt Kosh	Digitized Blood Bank Network
My Health Record	Personal Health Records
Health Wellness Calendar	Health Calendar
Family Planning (humdo.nhp.gov.in)	Family Planning and Information Portal
M-Diabetes	
Telemedicine	National Medical College Network
	National Telemedicine Network
	Use of Space Technology for Telemedicine
Health Kiosk System	A Small Machine in a Public Place to Disburse Health Information

TABLE 6.3
Indicative List of Telemedicine Firms in India

Name of the Firm	Types of Services Provided
Mfine [28]	Health Checkups
	Physician Consultations
	Corporate Health Services
1mg [29]	Physician Consultations
	Online Pharmacy
Practo [30]	Online Physician Appointments
	Physician Consultations
	Online Pharmacy
	Diagnostics
Lybrate [31]	Physician Consultations
	Online Pharmacy
Medlife [32]	Physician Consultations
	Online Pharmacy
Portea Medical [33]	Physician Consultations
	Online Pharmacy
	Medical Equipment
Apollo Pharmacy [34]	Physician Consultations
Apollo Telemedicine Networking Foundation (ATNF) [35]	Sustainable Community Healthcare and Telemedicine Systems
Apollo 247 [36]	Online Pharmacy
	Medical Equipment
World Health Partners [37]	Telemedicine Solution
	Call Center
Telerad Providers [38]	Telemedicine Solution
Rijuven India [39]	Telemedicine Solution
	Medical Equipment
Rank Tech solutions [40]	Telemedicine Solution
	Medical Equipment
Kria Healthcare [41]	Telemedicine Solution
	Trainings
Medi Buddy [42]	Physician Consultations
	Online Pharmacy
	Health Checkups
	Health Insurance
Medlife [43]	Online Pharmacy
	Physician Consultations
	Diagnostics
DocsApp [44]	Online Pharmacy
	Physician Consultations
	Diagnostics

(Continued)

TABLE 6.3 (*Continued*)
Indicative List of Telemedicine Firms in India

Name of the Firm	Types of Services Provided
Tata Health [45]	Online Pharmacy
	Physician Consultations
	Diagnostics
	Instant Doctor Consultation 24/7
PharmEasy [46]	Online Pharmacy
	Diagnostics
Amazon India Health Care Store [47]	Online Pharmacy
Smart Medics [48]	Physician Consultations
BookMEDS [49]	Online Pharmacy
MedPlusMart [50]	Online Pharmacy
Yodawy [51]	Online Pharmacy

6.5.2 ANALYSIS II

In the era of the knowledge economy and the Fourth Industrial Revolution (4IR) eco-system, it is imperative to devise meaningful conjugation between health informatics and telemedicine in India. The synergetic assimilation of virtual health resources would essentially improvise e-health strategy across the country. The e-health strategy is a more reliable and cost-effective framework to penetrate the remote inaccessible regions of rural India. The COVID-19 pandemic has taught a great lesson to all of us because it portrays that the best health infrastructural system of the world may crumble to address any severe pandemic or humanitarian crisis. We need to understand 68% of the population stays in rural India.

Figure 6.1 describes how an organized smart e-health platform facilitates the rural patient to access quality medical treatment blended with physical and virtual modes of treatments.

This is the algorithm showing smart e-health framework for rural India:

1. Patients in rural India generally visit any qualified doctor working at nearby health centers, PHC, etc.
2. If the patients do not recover, they may be referred by the local doctor to visit a super-specialist generally available in the city or district headquarters.
3. If it is not feasible, the traditional rural health infrastructure comes to a deadlock, which can be improvised with the advent of appropriate e-health strategy.
4. Under this framework, the local e-health facility station needs to be established in the existing sub-center, PHC, and CHC where a qualified junior medical officer would interact with the critical patient and liaison with a

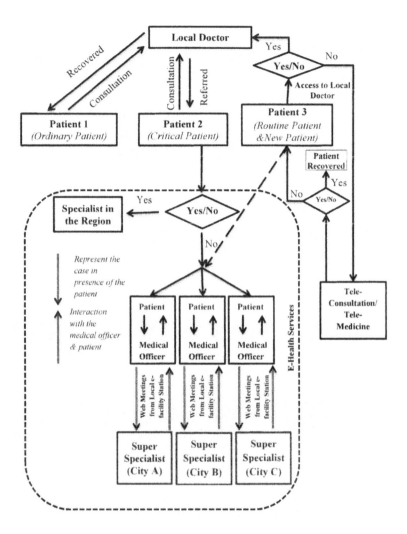

FIGURE 6.1   Improvised rural e-health framework along with blended services.

super-specialty doctor across the country for enabling virtual consultation
with the critical patients.

5. This would help the critical patient in the rural area to avail of the best possible consultation with almost no mobility for patients and escort.

6. If it becomes impossible to get basic consultation form the local doctor for
   any routine or new patient, telemedicine or tele-consultation would have to
   be deployed so that the patient could get immediate consultation and relief.

7. If the treatment through telemedicine or tele-consultation fails to treat the
   patient's illness, they may go to a rural e-health facility station for a super-
   specialist consultation through the web or videoconferencing.

This comprehensive framework would serve health-related challenges in rural areas in many ways:

- Firstly the frame work would minimize the delay for availing treatment by the patient.
- Secondly the model would also ensure availing super-specialty advice with almost zero mobility of patient and escort, if required.

The other collateral benefits of the model are depicted below:

- The rural people can save on nonmedical expenditures and minimize the loss of economy by saving man-days.
- This model would address not only the healthcare issues, but also it would reduce morbidity and thus reinforce the rural economy.
- This framework doesn't require exorbitant cost or huge investment for both the service provider and seekers.

6.5.3 ANALYSIS III

In the healthcare system, it is believed that "Prevention is better than cure," which has been seen in the recent pandemic. The global health experts have been emphasizing social distancing, lockdowns, quarantine, etc., which are of a precautionary or preventive measure. The global fraternity begins to learn the paradox of herd immunity. The lack of a specific medicine or vaccine for treating COVID-19 patients has been popularized in the doctor's community for improving body immunity. The COVID victims are prescribed largely vitamins, zinc supplement drugs. In recent years there has been a growing trend of practicing yoga worldwide to make synergy between body and mind. On the other hand, mental illness has emerged as the next pandemic spread throughout the world. WHO data suggests that globally 264 million people are affected by depression, 45 million suffer from bipolar disorder, 20 million people suffer from schizophrenia, and 50 million suffer from dementia [52].

Immunity centers will be located within the highest possible reach in rural areas. It could be established in all existing sub-centers, PHC, and CHC. The infrastructure of PRI can be utilized to establish immunity centers as shown in Figure 6.2. Apart from that, the immunity center may be established in schools, colleges, and universities across the country. The immunity center can also be awarded to reputed health nongovernmental organizations (NGOs) in the region with appropriate terms and conditions. Immunity centers would be operationalized using existing unused resources like specific school buildings in the evenings or on holidays with no additional cost or with minimal variable expenditure.

Every immunity center ideally should have a qualified nutritionist, trained yoga expert, and a dedicated psychiatrist or physiologist. This team would necessarily be monitored and mentored by a qualified medical doctor. The immunity center would provide consultation, mentoring, and guidance to every individual with respect to three broad dimensions: nutrition, yoga therapy, and mental health.

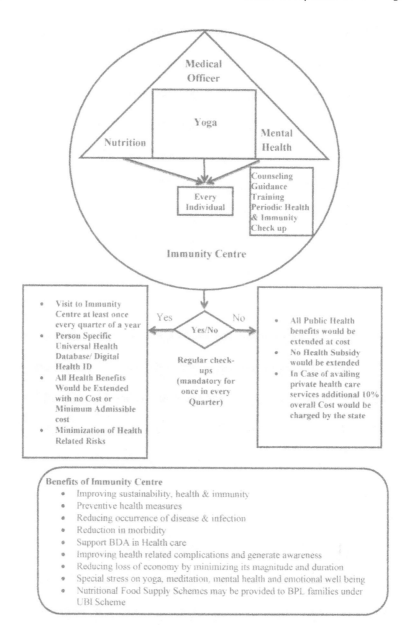

FIGURE 6.2 Perceptual immunity center framework.

The expected benefits may include:

1. If the state can adopt partial modes of universal basic income (UBI) [53] to protect the health and livelihoods of the people below the poverty level. Under this UBI scheme, if adopted by the state then the immunity center

would be able to provide food and nutrition supplements to the needy and destitute section of the society.

2. To bring rapid transformation to the lives of rural people, the state can impose that every citizen shall have to report and receive consultation once in a quarter, failing which the state-sponsored health schemes may not be extended free of cost or at a marginal expense. Even the defaulter may be imposed up to 10% of charges on overall expenditure incurred in private hospitals. All the information regarding a regular visit to the respective immunity center would be recorded in the person-specific Universal Health Database or Health ID.

3. This regular monitoring of health would generate multiple data sets that can solve future health hazards using appropriate big data analytics strategies for solving critical and complex ailments.

4. The proposed formation for the immunity center would be able to develop a culture of generating health awareness, building individual immunity followed by drawing attention to yoga, freehand excesses, etc.

5. The immunity center would also recognize the growing importance of fostering mental health followed by appropriate psychosomatic intervention so that the emerging pandemic can be registered to a greater extent.

6. The proposed immunity center may act as an e-health facility station to promote a virtual mode of super-specialty consultation and treatment.

6.6 CONCLUSION

This chapter show cased the existing infrastructure and practices of e-health initiatives in India. It recommends appropriate policy interventions that may be helpful for policymakers, health professionals, and administrators of the country. The research study also indicates the path for manifesting target-oriented achievements for the health sector in consonance with the dictum of UNSDGs.

REFERENCES

1. Chakrabarty, A., and Das, U. S. (2020). Universal Health Database in India: Emergence, Feasibility and Multiplier Effects. In A. Mishra, G. Suseendran, & T.-N. Phung (Eds.), *Soft Computing Applications and Techniques in Healthcare*. CRC Press Taylor & Francis Group.
2. Chakrabarty, A., and Das, U.S. (2020). Big Data Analytics in Excelling Health Care: Achievement and Challenges in India. In P. Tanwar, V. Jain, C.-M. Liu, & V. Goyal (Eds.), *Big Data Analytics and Intelligence: A Perspective for Health Care*, pp. 55–74. Emerald Publishing Limited. https://doi.org/10.1108/978-1-83909-099-820201008
3. CBHI, MHFW, GoI. 2019. *National Health Profile (NHP) of India- 2019 : Ministry of Health and Family Welfare*. http://cbhidghs.nic.in/showfile.php?lid=1147.
4. "EHealth Solutions – Siemens Healthineers India." n.d. Accessed October 27, 2020. https://www.siemens-healthineers.com/en-in/digital-health-solutions/digital-solutions-overview/patient-engagement-solutions/e-health-solutions.
5. WHO. 2018. "Classification of Digital Health Interventions." *World Health Organization* 14 (4): 12–18. http://who.int/reproductivehealth/topics/mhealth/en/.

6. World Health Organization. 2019. *WHO Guideline: Recommendations on Digital Interventions for Health System Strengthening. Food and Nutrition Bulletin.* Vol. 2. World Health Organization. http://www.who.int/reproductivehealth/publications/digital-interventions-health-system-strengthening/en/.

7. FICCI. 2016. "Health Information Technology- A New Initiative." http://ficci.in/spdocument/20101/Status-Paper-Health-IT.PDF.

8. MoHFW. 2006. "National Health Systems Resource Centre, MoHFW, Government of India." 2006. http://www.nhsrcindia.org/health-informatics.

9. "How Technology Is Changing Health Care in India - Knowledge@Wharton." 2020. 2020. https://knowledge.wharton.upenn.edu/article/technology-changing-health-care-india/.

10. Jnr, Bokolo Anthony. "Use of telemedicine and virtual care for remote treatment in response to COVID-19 pandemic." *Journal of Medical Systems* 44, no. 7 (2020): 1–9.

11. Kamal, S. A., Shafiq, M., and Kakria, P. (2020). Investigating acceptance of telemedicine services through an extended technology acceptance model (TAM). *Technology in Society 60*, 101212.

12. Dorsey, E. Ray, and Eric J. Topol. "Telemedicine 2020 and the next decade." *The Lancet* 395, no. 10227 (2020): 859

13. Mann, Devin M., Ji Chen, Rumi Chunara, Paul A. Testa, and Oded Nov. "COVID-19 transforms health care through telemedicine: evidence from the field." *Journal of the American Medical Informatics Association* (2020).

14. Calton, Brook, Nauzley Abedini, and Michael Fratkin. "Telemedicine in the time of coronavirus." *Journal of Pain and Symptom Management* (2020).

15. Chenthara, Shekha, Khandakar Ahmed, Hua Wang, and Frank Whittaker. "Security and privacy-preserving challenges of e-health solutions in cloud computing." *IEEE access* 7 (2019): 74361–74382.

16. Zhang, Aiqing, and Xiaodong Lin. "Towards secure and privacy-preserving data sharing in e-health systems via consortium blockchain." *Journal of medical systems* 42, no. 8 (2018): 140.

17. Lin, Di, Yu Tang, Fabrice Labeau, Yuanzhe Yao, Muhammed Imran, and Athanasios V. Vasilakos. "Internet of vehicles for E-health applications: A potential game for optimal network capacity." *IEEE Systems Journal* 11, no. 3 (2015): 1888–1896.

18. Gallagher, Joe, John O'Donoghue, and Josip Car. "Managing immune diseases in the smartphone era: how have apps impacted disease management and their future?." (2015): 431–433.

19. Li, JunHua, Pradeep Ray, Holly Seale, and Raina MacIntyre. "An E-Health readiness assessment framework for public health services–Pandemic perspective." In *2012 45th Hawaii International Conference on System Sciences*, pp. 2800–2809. IEEE, 2012.

20. Bazzani, Marco, Davide Conzon, Andrea Scalera, Maurizio A. Spirito, and Claudia Irene Trainito. "Enabling the IoT paradigm in e-health solutions through the VIRTUS middleware." In *2012 IEEE 11th International Conference on Trust, Security and Privacy in Computing and Communications*, pp. 1954–1959. IEEE, 2012.

21. El Khaddar, Mehdia Ajana, Hamid Harroud, Mohammed Boulmalf, Mohammed Elkoutbi, and Ahmed Habbani. "Emerging wireless technologies in e-health trends, challenges, and framework design issues." In *2012 International Conference on Multimedia Computing and Systems*, pp. 440–445. IEEE, 2012.

22. Liu, Caiming, Minhua Guo, Lingxi Peng, Jing Guo, Shu Yang, and Jinquan Zeng. "Artificial immunity-based model for information system security risk evaluation."

In *2010 International Conference on E-Health Networking Digital Ecosystems and Technologies (EDT)*, vol. 1, pp. 39–42. IEEE, 2010.

23. Li, JunHua, and Pradeep Ray. "Applications of E-Health for pandemic management." In *The 12th IEEE International Conference on e-Health Networking, Applications and Services*, pp. 391–398. IEEE, 2010.
24. Kluge, Eike-Henner W. "Secure e-health: managing risks to patient health data." *International Journal of Medical Informatics* 76, no. 5-6 (2007): 402–406.
25. National Sample Survey Organization. "Health in India (NSS 71st round)." (2014).
26. "India: Attitude towards Using Telemedicine 2018 | Statista." n.d. Accessed October 27, 2020. https://www.statista.com/statistics/917308/india-attitude-towards-using-telemedicine/.
27. "National Health Portal of India, Gateway to Authentic Health Information." n.d. Accessed October 27, 2020. https://www.nhp.gov.in/.
28. "MFine." n.d. Accessed October 27, 2020. https://www.mfine.co/.
29. 1mg. 2020. "Online Pharmacy India | Buy Medicines from India's Trusted Medicine Store." 2020. https://www.1mg.com/.
30. "Practo | Video Consultation with Doctors, Book Doctor Appointments, Order Medicine, Diagnostic Tests." n.d. Accessed October 27, 2020. https://www.practo.com/.
31. "Online Doctor | Ask A Doctor Online | Lybrate." n.d. Accessed October 27, 2020. https://www.lybrate.com/.
32. "Medlife: Online Medicine - India's Largest Online Pharmacy Store." n.d. Accessed October 27, 2020. https://www.medlife.com/.
33. "Healthcare at Home: India's #1 Home Care Services." n.d. Accessed October 27, 2020. https://www.portea.com/.
34. "A Promise beyond Prescriptions Buy Medicines Online | Order Online Medicines | Apollo Pharmacy." n.d. Accessed October 27, 2020. https://www.apollopharmacy.in/.
35. "Apollo Telemedicine Networking Foundation (ATNF) | The Center for Health Market Innovations." n.d. Accessed October 27, 2020. https://healthmarketinnovations.org/program/apollo-telemedicine-networking-foundation-atnf.
36. "Apollo 247 - Online Doctor Consultation & Online Medicines, Apollo Pharmacies Near Me." n.d. Accessed October 27, 2020. https://www.apollo247.com/.
37. Chavali, Annapurna. 2011. "World Health Partners," 1–22. https://worldhealthpartners.org/.
38. "Teleradiology Providers New Delhi, Teleradiology Providers India." n.d. Accessed October 27, 2020. http://teleradproviders.com/.
39. "Rijuven India | Smart Telemedicine." n.d. Accessed October 27, 2020. https://rijuvenindia.com/home.html.
40. "Innovator of World's First Video Banking Solution – RankTech Solutions." n.d. Accessed October 27, 2020. https://www.ranktechsolutions.com/.
41. "KRIA – Health | Care | Happiness " n.d. Accessed October 27, 2020. http://www.kriahealth.com/.
42. "Book Health Checks, Lab Tests, Online Medicine & Doctor Consultation | MediBuddy." n.d. Accessed October 27, 2020. https://www.medibuddy.in/.
43. Medlife. 2018. "Medlife: Online Medicine - India's Largest Online Pharmacy Store." 2018. https://www.medlife.com/.
44. "DocsApp - Online Doctor Consultation App, Consult Doctor on Chat & Call – Dr 24x7." n.d. Accessed October 27, 2020. https://www.docsapp.in/.
45. "Consult a Doctor or Book an Appointment Online | Tata Health." n.d. Accessed October 27, 2020. https://www.tatahealth.com/.
46. "PharmEasy: Online Pharmacy & Medical Store in India | 5M+ Customers." n.d. Accessed October 27, 2020. https://pharmeasy.in/.

47. "Buy Amazon Basics Food Feeder, Small Online at Low Prices in India - Amazon. In." n.d. Accessed October 27, 2020. https://www.amazon.in/health-and-personal-care/b?ie=UTF8&node=1350384031.
48. "SmartMedics| Talk To a Doctor on Phone or Consult Online in 15 Minutes." n.d. Accessed October 27, 2020. https://www.smartmedics.co/.
49. "BookMeds." n.d. Accessed October 27, 2020. http://www.bookmeds.com/sitemap.
50. "Online Pharmacy Store in India. Best Value on Medicines - MedPlusMart." n.d. Accessed October 27, 2020. https://www.medplusmart.com/.
51. "Your Pharmacy Benefits." n.d. Accessed October 27, 2020. https://www.yodawy.com/.
52. "MENTAL DISORDERS." *Medical Journal of Australia* 2, no. 5 (1930): 159–60. https://doi.org/10.5694/j.1326-5377.1930.tb41371.x.
53. Chakrabarty, Arindam. "Universal Basic Income At The Bottom Of The Pyramid: Achieving Sdgs Through Financial Inclusion In India." *Administrative Development' A Journal of HIPA, Shimla'* 6, no. 1 (2019): 119–141.

7 Predicting Medical Procedures from Diagnostic Sequences Using Neural Machine Translation

Siddhanth Pillay and Sowmya Kamath S

CONTENTS

7.1 INTRODUCTION

Over the past decade, advances in digitization have resulted in unprecedented efforts to digitize patient records and make them available for consumption as electronic health records (EHRs), which has been highly successful, especially in developed countries like the United States, Canada and Australia [1]. EHRs capture the complete medical history and the current condition of a patient with detailed information on symptoms, diseases, medications, allergies, hospital admissions, etc. Hence, EHRs are a rich source of clinical information that can be put to use for the betterment of healthcare systems and in improving care delivery and management, specifically by applying technologies like data analytics and machine learning. Envisioned applications like predictive modeling, preventive modeling, automatic information/ concept extraction, recommendations for doctors and patients, etc. can positively impact the way health care is delivered and managed.

Consequently, a lot of data, hitherto difficult to procure, process, and analyze, has now become easily available for gaining insightful information. Several studies have proven that when harnessed effectively, this data can give beneficial and tangible insights into patients' profile modeling and their treatments, and also help minimize escalating healthcare costs [2, 3]. One such category of data that is readily available is diagnoses codes and the corresponding procedure codes in a standardized format like ICD-9[1] codes. The question that then arises is - how can we effectively use the data in order to identify patterns? One possible usage is using the diagnoses codes for higher-level purposes like predicting treatment procedures. As diagnosis codes are nothing but manifestations of the underlying disease, and are recorded along with the medical techniques that need to be performed to remedy the disease, diagnosis codes and procedure codes are conceptually related. This intuition can be easily captured using neural networks over large-scale patient data and can be used to develop effective clinical decision support systems (CDSSs) that can aid medical personnel in optimizing patient care and cutting down healthcare costs.

Given a set of diagnoses for a patient who presents with no prior history, is it possible to predict the set of medical procedures to be performed on the patient as accurately as possible? Toward achieving this target, we view the problem as essentially one of neural machine translation [4], where the `translation' task deals with the purpose of mapping the diagnosis to the procedure space. A dataset containing the diagnoses codes of patients and the corresponding treatment is obtained from a large-scale open dataset, and a deep neural model is trained to perform the prediction task. As LSTMs (Long Short-Term Memory networks) [5] are most widely used for processing sequential data, the proposed model leverages LSTMs for this purpose. Experimental validation of the proposed approach showed promising results, which are presented in the form of BLEU scores because this is essentially a translation task. The advantage of the approach is that it provides an insight into the relation between the diagnosis codes and their relation with the procedure codes, which can be leveraged for treatment recommendation. To the best of our knowledge, our approach is unique and has not been tried earlier.

The work presented in this chapter is organized as follows. Section 7.2 covers a brief discussion of the existing works and approaches for addressing the problem of interest. In Section 7.3, we describe in detail the architecture and processes designed as part of the proposed approach. Section 7.4 discusses the observed experimental results and justification relating to the observations, followed by a conclusion and potential directions for future research.

7.2 RELATED WORK

Extensive research in the field of designing intelligent applications for healthcare data has been made possible primarily due to the digitization and subsequent availability of large-scale patient data [6]. Of these, disease prediction is a major research challenge that has generated significant research interest. Miotto et al. [7] employed an architecture comprising three sets of denoising autoencoders to identify the characteristics and interdependencies in large-scale structured health records, using which, they are able to predict a certain set of diagnoses. This was one of the first approaches that effectively demonstrated the significant edge demonstrated by deep neural models over any other features' selection techniques. They demonstrated the

---

[1] CDC, "International Classification of Diseases, 9th revision," https://www.cdc.gov/nchs/icd/icd9cm.htm.

effectiveness of vectorized representations of patient-specific information, through their model named *Deep Patient*, which uses unsupervised feature learning and denoising autoencoders to learn disease nuances iteratively.

Researchers have also focused on processing the temporal sequence of occurrence of events in patient data. Lipton et al. [8] used an LSTM model to identify interesting trends in a set of clinical measurements over time, and classified them to predict potential diagnoses. They made the use of a sequence of LSTMs along with a fully connected layer for performing multilabel classification. They perform sequential target replication at every sequence step and then calculate a comprehensive loss, which is a combination of the log loss at each output node, as well as the local loss at each sequence step. Razavian et al. [9] applied both LSTM and convolutional neural network (CNN) models for essentially the same task. They use a sliding window mechanism on the longitudinal measurement of lab tests to predict the onset of diseases. An obvious problem that arises is that the length of the sliding window has to be heuristically tuned. Aczon et al. [10] used temporal laboratory measurements for predicting mortality risk of patients treated in a pediatric intensive care unit. The model takes in a sequence of information at a particular time step, and produces a mortality risk prediction at a specified future time. At each time step, a comprehensive sequence of data is entered into the model, which consists of physiological observations, lab results, interventions, and drugs.

Rajkomar et al. [11] and Gangavarapu et al. [12, 13] applied sequential data processing techniques to a large patient database to predict concepts like mortality when admitted to the hospital, 30-day planned readmission, prolonged length of stay as well as patient-specific final diagnoses. They were able to achieve high accuracy in their predictions. They also developed a neural network attribution mechanism that demonstrated how clinicians can gain some amount of transparency in the predictions made by the system. In Doctor AI [14], the temporal events associated with a patient are inputted, where each temporal event can be a multi-hot label vector of diagnosis codes, procedure codes or medication codes along with time duration until the next visit, and the output is a prediction of the diagnoses codes and the medication codes as well as the duration until the next visit. Effectively, the RNN model *learns* the patient status at any given point of time and makes future predictions based on that status. Though this is efficient, it depends on the accuracy of learning the patient status. In contrast, we model our approach such that there is no need to learn *patient status*, the model uses the concept of neural machine translation, which enables us to translate a set of diagnoses from the diagnoses space to the procedure space.

LSTMs are RNN architectures that have been applied to a variety of problems in the areas of natural language processing (NLP) and analytics. Most LSTMs also make use of forget gates [15] and peephole connections [16] that enhance their prediction accuracy. Hochreiter et al. [5] modeled LSTM-based architectures, specifically for sequence learning. Lipton et al. [17] wrote a detailed survey on the application of RNNs for sequence learning. De et al. [18] detailed the different steps in computing a language model using RNNs and surveyed all the applications in an elaborate manner. Sutskever et al. [4] used an LSTM-based model for the purpose of neural machine translation, in which they performed English to French translation tasks quite successfully. Kalchbrenner et al. [19] used CNNs for similar purposes, but they failed to maintain the ordering of the words, which was a major drawback.

More recently, research on automatic code assignment has been attempted by modeling the unstructured clinical text [13, 20–26], thus, exploring the richness of patient-specific information in such free text. Most works have utilized the standard, openly available Medical Information Mart for Intensive Care III (MIMIC-III) database [27], comprising over 40,000 patients' data. Recently, a dataset containing expert-annotated data was released as part of the CLEF eHealth 2020 challenge [28], which allows researchers to validate the interpretation capabilities of their model's decision-making processes. Cho et al. [29] made use of a novel RNN architecture that primarily targeted the integration of neural networks into a statistical machine translation (SMT) system by mapping sentences into a vector and back. These two RNNs are jointly trained on the conditional probability of the target sequence, when the source sequence is given. In contrast to these approaches, we adopt a singularly novel methodology of treating the problem of disease code mapping to diagnosis code prediction as a neural machine translation task, using LSTMs modeled as an encoder-decoder network. This approach has not been attempted before as per our extensive literature review, and we show experimentally that it is well-suited for the task at hand.

7.3 PROPOSED APPROACH

The motivation behind designing a translation-based approach for predicting medical diagnoses stemmed from the challenge of disambiguating between multiple procedure codes for a particular disease. Here, one can argue that any given diagnosis can have only one particular associated procedure (or a list of procedures), so there is no need for any "prediction." However, it is also true that each diagnosis can have multiple procedure codes, and they may share a conjunctive or disjunctive relation, thus complicating the process. Also, a Diagnosis A can inhibit the performance of a procedure associated with Diagnosis B if both Diagnoses A and B occur together, and hence, some alternate procedure may have to be performed for Diagnosis B. This would require a hard-coded rule-based system, the development of which becomes a highly cumbersome process and may fail to account for all possibilities. Thus, intuitively, a rule-based system is infeasible for most practical scenarios. The essence of a translation-based approach is that it can capture the semantic/latent concepts behind source and target sentences to understand meaning. As diagnosis codes and procedure codes for a particular hospital visit are conceptually linked by the underlying disease, we propose a translation-based approach to predict the procedure sequences. Figure 7.1 depicts the processes designed for the prediction task. We used a publicly available, open dataset called MIMIC-III [27], which provides information relating to patients admitted to critical care units. The patient data provided includes details on prescribed drugs, lab tests, unstructured notes written by doctors and nurses during patient care, ICD-9 codes detailing procedures and diagnosis, radiology reports, and so on. For the purpose of training, we make use of ICD-9 codes of the diagnoses and the associated procedures.

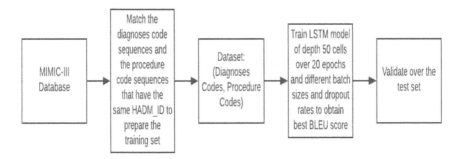

FIGURE 7.1 Preprocessing tasks performed to generate the training dataset.

We extracted the diagnoses codes and procedure codes for patients indexed by the patient ID (SUBJECT ID) and the hospital admission ID (HADM ID) from the dataset. A single patient could have multiple hospital visits, which means a unique SUBJECT ID can have multiple associated HADM IDs. However, a unique HADM ID has a single sequence of diagnosis code and procedure code, and so, we extract the corresponding diagnosis code and procedure code for use as training data. These codes are fed to an encoder LSTM model for training, while the decoder LSTM decodes these sequences to output the sequence of y_is, $i = 1, ..., T^1$ step by step, where each y_i represents a potential procedure code. Note that the lengths of the diagnoses codes and the procedure codes may be different, i.e., T and T^1 may be different. The proposed model is primarily an LSTM modeled as an encoder-decoder model (EDM), as depicted in Figure 7.2.

It consists of a memory cell, an input gate, an output gate and a forget gate. The rate of input data flow is managed by the input gate, while the function of the forget gate is to decide on the amount of information to be retained. The output gate is responsible for decisions on the effect of the cell value on the output activation of the LSTM unit. The EDM is governed according to various parameters computed as per Eq. (7.1) to (7.6). In Eq. (7.1), i represents the input layer, which determines what values are to be updated, followed by a tanh activation function These two are combined, as shown in Eq. (7.5), to generate an update to the current state. As shown in Eq. (7.2), f represents the forget gate layer, which is a sigmoid function. The

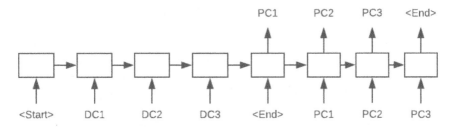

FIGURE 7.2 An example of the EDM at work. Here, a diagnosis code sequence is processed code by code by the encoder to generate the vector representation, which the decoder uses to generate procedure sequences.

previous state of the LSTM s_{t-1} and the current input x_t are given as input, and a value between 0 and 1 is generated (where, 0 intuitively represents *forget everything* and 1 represents *remember everything*), which is combined with the previous cell state c_{t-1}. Finally, the output o is calculated to represent the parts of the cell state that will be retained, which is combined with a tanh-filtered cell state to obtain the final state.

$$i = \sigma \left(x_t U^i + s_{t-1} W^i \right) \tag{7.1}$$

$$f = \sigma \left(x_t U^f + s_{t-1} W^f \right) \tag{7.2}$$

$$o = \sigma \left(x_t U^o + s_{t-1} W^o \right) \tag{7.3}$$

$$g = tanh \left(x_t U^g + s_{t-1} W^g \right) \tag{7.4}$$

$$c = c_{t-1} \odot f + g \odot i \tag{7.5}$$

$$s_t = tanh(c_t) \odot o \tag{7.6}$$

In the proposed model, two LSTMs are employed, one for encoding and the other for decoding, in contrast to Sutskever et al.'s model [4]. The encoder LSTM accepts the sequence of diagnoses codes as input and then generates a fixed-size vector that effectively represents the sequence. Beginning at the *<START>* tag, the encoder LSTM processes each term of the sequence in a step-wise manner and at each step, the internal cell state is also updated. After processing the input sequence, i.e., when the *<END>* tag is encountered, the internal cell state of the encoder LSTM is the fixed-size vector representation of the input sequence. This vector is then fed into the decoder LSTM, which maps it to the target sequence, thus making this a translation task from diagnoses space to procedure space. The decoder LSTM predicts the target sequence by effectively estimating the conditional probability $P\,(j_1, j_2, ..., j_{T^1}\,|\,i_1, i_2, ..., i_T)$, where $(i_1, i_2, ..., i_T)$ is the input sequence of length T and $(j_1, j_2, ..., j_{T^1})$ is the target sequence of length T^1. Note that the lengths of the two sequences, T and T^1 need not be the same. The decoder LSTM calculates the conditional probability by first calculating the fixed-size vector v, after which the probability of $j_1, j_2, ..., j_{T^1}$ is calculated as per Eq. (7.8).

Finally, to represent all the index terms in the vocabulary for each $P\,(j_t\,|\,v, j_1, ..., j_{t-1})$, a softmax function is used. The LSTM formulation used in the proposed model is adapted from the model proposed by Graves et al. [30]. Here, the beginning of each sequence is marked with a *<START>* token and the ending is marked with an *<END>* token, to enable the model to define the probability distribution even over sequences having variable length. At each intermediate step t, the probability is calculated by taking into account v along with all previous outputs $j_1, j_2, ..., j_{t-1}$. The probability of the entire sequence is given by Eq. (7.7), and is a function of the sequence's individual term probabilities, while assuming a mutual exclusion between the individual terms.

$$T^1 \; p\left(j_1, \, j_2, \, ..., j_{T^1} \mid i_1, \, i_2, \, ..., \, i_T\right) = \prod P\left(j_t \mid v, j_1, \, ..., \, j_{t-1}\right) t = 1 \qquad (7.7)$$

7.4 EXPERIMENTS AND RESULTS

The experimental validation of the proposed neural translation-based approach is described in detail here. For the experiments, we used the patient data from MIMIC-III, which provides extensive structured and unstructured medical data relating to the hospital admissions of about 38,597 patients who underwent critical care procedures at a reputed hospital over a period of 12 years. A total of 49,785 hospital admission episodes of these patients are made available for analytics. Owing to the relatively modest size of the available dataset, we used 95% of the patient data for training and 5% for testing. The dataset contained around 6,817 distinct ICD-9 diagnostic codes in the input vocabulary and 2,011 distinct ICD-9 procedure codes in the output vocabulary. Since our vocabulary is fixed, we used a fixed vector representation in order to represent the vocabulary words. The EDM was trained for 20 epochs and then evaluated for prediction performance. The number of hidden layers was limited to 50 for faster computational purposes. However, we are currently experimenting with a higher number of hidden layers so as to achieve maximum possible expressive power. During various experimental runs, we determined that the Adam optimizer worked best for our dataset. We chose a learning rate of 0.001, $\beta_1 = 0.9$ and $\beta_2 = 0.999$. Further, we also found that a batch size of 64 worked best when combined with a dropout rate of 30% on the input connections to the LSTMs, which reduced the overfitting problem significantly. The model was trained so that the log probability of translation T having a source sentence S is maximized, both belonging to the training set S (Eq. 7.8). By finding the most likely translations produced by the LSTM, we produce the translations as per Eq. (7.9).

$$Objective = 1/|S| \; \Sigma \; log \; p\left(T|S\right) (T,S) \in S \qquad (7.8)$$

$$T^\wedge = arg \; max \; p\left(T|S\right)_T \qquad (7.9)$$

As a baseline, a fully connected model having two hidden layers was also trained. The input data is a multi-hot vector $x = \{0, 1\}^p$ where the value was set to 1 if the diagnosis code with that particular index is present in the sequence and the output is a multi-label vector. Although the model performed well in the initial few stages, within just 2 to 3 epochs, it started overfitting drastically as expected. We experimented to overcome this problem by applying a wide range of dropouts, which did not help much either. Based on these observations, we concluded that the proposed translation-based approach potentially works better for problems of sequence generation of procedures. For evaluating the proposed approach, we chose a set of suitable metrics as per the objectives defined. Since the task is essentially one of translation,

TABLE 7.1

Performance of EDM Model of Depth 50 on Batches of varying sizes, trained for 20 Epochs, without Dropout

Batch Size	BLEU Score
16	0.4163
32	0.4439
64	0.4489

we chose the BLEU score as a suitable metric. BLEU is a popular metric that is used to compare a candidate translation with given reference translation(s). It is essentially a modified version of precision, where, for a given n-gram in the candidate translation, the maximum possible occurrences of that n-gram across the given set of reference translations is computed. This process is repeated for all possible n-grams and finally, the ratio of the sum of maximal occurrences to the number of unique n-grams in the candidate translation is computed to obtain the BLEU score. In our experiments, we considered $n = 1$, because greater importance is attached to obtaining the right procedure code, rather than the order in which procedure codes are obtained. Also, each translation has only one reference sentence, in our case. We calculated the sentence-level BLEU scores using Python's NLTK package, and the average of the scores is presented as the final score here.

To evaluate the EDM model, we trained it by accommodating for batch sizes of 16, 32, and 64, where the depth of the LSTM model was 50 cells and the dropout was 0. The observed scores are tabulated in Table 7.1, which shows that the best performance was obtained with a batch size of 64 since the decrease in loss was the steepest in this case (graphically illustrated in Figure 7.3).

This is further reinforced while analyzing the obtained BLEU score, which was the highest for batch size 64 (0.4489), as can be observed from the results depicted in Table 7.1. Since the size of our dataset is relatively small, we also had to account for the possibility of over fitting. Hence, we trained the model with the optimal batch size of 64 at different dropout rates, for further analysis. As can be observed from

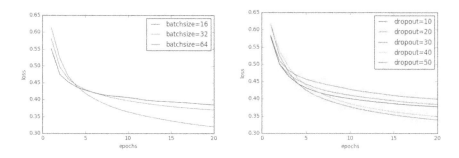

FIGURE 7.3 Evaluation of loss on the training set.

TABLE 7.2

Performance of EDM Model of Depth 50, trained for 20 Epochs and with optimal Batch size of 64, at different Dropout rates

Dropout Rate	BLEU Score
0.1	0.4094
0.2	0.4356
0.3	0.4284
0.4	0.4303
0.5	0.3814

Table 7.2, the steepest decrease in loss occurs in the case of dropout rates of 30%, followed by that of 40%, which are both very close. This is also validated by the BLEU scores in Table 7.2, where the scores of the two dropout rates are very close to each other. The BLEU scores that were obtained after tuning for the best parameters, proved that there is certain merit to the idea of a translation-based approach to the task of procedure prediction. Thus, our hypothesis is validated and calls for a further in-depth analysis into this approach.

7.5 CONCLUSION AND FUTURE WORK

In this work, a neural translation-based approach for predicting medical procedures from the diagnoses sequences using an LSTM model was presented. Our model is built on encoder and decoder LSTMs, which translate the input diagnostic codes to the procedure space, for applications like the input diagnostic codes to the procedure space, for applications like treatment/procedure recommendation. The experimental results presented in the form of BLEU scores were promising. Given the encouraging results, we are currently working on an ensemble model composed of two separate LSTMs trained on the diagnostic codes to use their combined outputs to generate procedure sequences. Although this work is in the preliminary stages, it has shown promising results. We also intend to address one of the current model's drawbacks, i.e., incorrectly predicted codes. Although the current performance of the model in terms of number of incorrect predictions is quite low statistically speaking, we are working toward further improving our model in order to eliminate the issue of any false negative predictions to optimize the proposed model for real-world deployment.

ACKNOWLEDGMENT

This research work was supported by the infrastructure provided by the Science and Engineering Research Board, Department of Science and Technology, Government of India, through the Early Career Research Grant (No. ECR/2017/001056) to the second author.

REFERENCES

1. EMC Digital Universe: The digital universe – Driving data growth in healthcare. Online: https://www.cycloneinteractive.com/cyclone/assets/File/digital-universe-healthcare-vertical-report-ar.pdf [Accessed 30-Aug-2020]
2. Bates, D.W., Saria, S., Ohno-Machado, L., Shah, A., Escobar, G.: Big data in healthcare: using analytics to identify and manage high-risk and high-cost patients. Health Affairs **33**(7) (2014) 1123–1131
3. Krumholz, H.M.: Big data and new knowledge in medicine: the thinking, training, and tools needed for a learning health system. Health Affairs **33**(7) (2014)
4. Sutskever, I., Vinyals, O., Le, Q.V.: Sequence to sequence learning with neural networks. In: Advances in Neural Information Processing Systems. (2014) 3104–3112
5. Hochreiter, S., Schmidhuber, J.: Long short-term memory. Neural Computation **9**(8) (1997) 1735–1780
6. Adler-Milstein, J., DesRoches, C.M., Kralovec, P., Foster, G., et al.: Electronic health record adoption in US hospitals: progress continues, but challenges persist. Health Affairs **34**(12) (2015) 2174–2180
7. Miotto, R., Li, L., Kidd, B.A., Dudley, J.T.: Deep patient: an unsupervised representation to predict the future of patients from the electronic health records. Scientific Reports **6** (2016) 26094
8. Lipton, Z.C., Kale, D.C., Elkan, C., Wetzel, R.: Learning to diagnose with lstm recurrent neural networks. arXiv preprint arXiv:1511.03677 (2015)
9. Razavian, N., Marcus, J., Sontag, D.: Multi-task prediction of disease onsets from longitudinal laboratory tests. In: Machine Learning for Healthcare Conference. (2016) 73–100
10. Aczon, M., Ledbetter, D., Ho, L., Gunny, A., Flynn, A., Williams, J., Wetzel, R.: Dynamic mortality risk predictions in pediatric critical care using recurrent neural networks. arXiv preprint arXiv:1701.06675 (2017)
11. Rajkomar, A., Oren, E., Chen, K., Dai, A.M., Hajaj, N., Hardt, M., Liu, P.J., Liu, X., Marcus, J., Sun, M., et al.: Scalable and accurate deep learning with electronic health records. npj Digital Medicine **1**(1) (2018) 18
12. Gangavarapu, T., Krishnan, G.S., Kamath, S.S., Jeganathan, J.: Farsight: Long-Term Disease Prediction Using Unstructured Clinical Nursing Notes. IEEE Transactions on Emerging Topics in Computing (2020)
13. Gangavarapu, T., Jayasimha, A., Krishnan, G.S., Kamath, S.S.: Predicting ICD-9 code groups with fuzzy similarity based supervised multi-label classification of unstructured clinical nursing notes. Knowledge-Based Systems (2019) 105321
14. Choi, E., Bahadori, M.T., Schuetz, A., Stewart, W.F., Sun, J.: Doctor ai: Predicting clinical events via recurrent neural networks. In: Machine Learning for Healthcare Conference. (2016) 301–318
15. Gers, F.A., Schmidhuber, J., Cummins, F.: Learning to forget: Continual prediction with LSTM. (1999).
16. Gers, F.A., Schmidhuber, J.: Recurrent nets that time and count. In: IJCNN, IEEE (2000) 3189
17. Lipton, Z.C., Berkowitz, J., Elkan, C.: A critical review of recurrent neural networks for sequence learning. arXiv preprint, arXiv:1506.00019 (2015)
18. De Mulder, W., Bethard, S., Moens, M.F.: A survey on the application of recurrent neural networks to statistical language modeling. Computer Speech & Language **30**(1) (2015) 61–98
19. Kalchbrenner, N., Blunsom, P.: Recurrent continuous translation models. In: Proceedings of the 2013 Conference on Empirical Methods in Natural Language Processing. (2013) 1700–1709

20. Zeng, M., Li, M., Fei, Z., Yu, Y., Pan, Y., Wang, J.: Automatic ICD-9 coding via deep transfer learning. Neurocomputing **324** (2019) 43–50

21. Gangavarapu, T., Jayasimha, A., Krishnan, G.S, Kamath S.S.: TAGS: Towards Automated Classification of Unstructured Clinical Nursing Notes. In: International Conference on Applications of Natural Language to Information Systems, Springer (2019) 195–207

22. Jayasimha, A., Gangavarapu, T., Kamath S, S., Krishnan, G.S.: Deep Neural Learning for Automated Diagnostic Code Group Prediction Using Unstructured Nursing Notes. In: Proceedings of the ACM India Joint International Conference on Data Science and Management of Data. CoDS-COMAD '20, New York, NY, USA, ACM (2020) 152–160

23. Gangavarapu, T., Krishnan, G.S., Kamath S.S.: Coherence-based modeling of clinical concepts inferred from heterogeneous clinical notes for ICU patient risk stratification. In: Proceedings of the 23rd Conference on Computational Natural Language Learning (CoNLL). (2019) 1012–1022

24. Xu, K., Lam, M., Pang, J., Gao, X., Band, C., Mathur, P., Papay, F., Khanna, A.K., Cywinski, J.B., Maheshwari, K., Xie, P., Xing, E.P.: Multimodal machine learning for automated icd coding. Proceedings of the 4th Machine Learning for Healthcare Conference. Volume 106 of Proceedings of Machine Learning Research., PMLR 197–215

25. Wang, S.M., Chang, Y.H., Kuo, L.C., Yun Nung Chen, F.L., Yu, F.Y., Chen, C.W.: Using deep learning for automated icd-10 classification from free text data. EJBI (01 2020) 1–10

26. Krishnan, G.S., Kamath, S.S.: Hybrid text feature modeling for disease group prediction using unstructured physician notes. In: International Conference on Computational Science, Springer (2020) 321–333

27. Johnson, A.E., Pollard, T.J., Shen, L., Li-wei, H.L., Feng, M., Ghassemi, M., Moody, B., Szolovits, P., Celi, L.A., Mark, R.G.: Mimic-iii, a freely accessible critical care database. Scientific Data **3** (2016) 160035

28. Goeuriot, L., Suominen, H., Kelly, L., Miranda-Escalada, A., Krallinger, M., Liu, Z., Pasi, G., Saez, G.G., Vi-viani, M., Xu, C.: Overview of the clef ehealth evaluation lab 2020. In: International Conference of the Cross-Language Evaluation Forum for European Languages, Springer (2020) 255–271

29. Cho, K., Van Merrienboer, B., Gulcehre, C., Bahdanau, D., Bougares, F., Schwenk, H., Bengio, Y.: Learning phrase representations using RNN encoder-decoder for statistical machine translation. arXiv preprint, arXiv:1406.1078 (2014)

30. Graves, A.: Generating sequences with recurrent neural networks. arXiv preprint, arXiv:1308.0850 (2013)

8 Sensing the Mood- Application of Machine Learning in Human Psychology Analysis and Cognitive Science

Ahona Ghosh and Sriparna Saha

CONTENTS

8.1   INTRODUCTION

The main goal of our modern age is a successful life. But we can attain this only if we stay relaxed and keep our mind healthy. Stress and depression make us sick, both mentally and physically, while peace and joy make our life healthy. Sometimes frustration rises so much that it completely destroys our lives and our careers. We need to identify it in time and control it properly so that we can live a healthy, happy, and fulfilling life. In these modern times, our scientists have advanced medical science using machine learning so much that now we have many methods or machines by

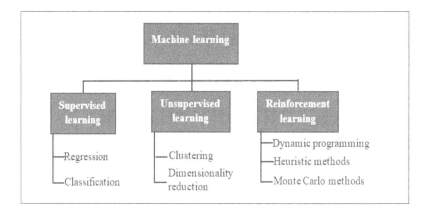

FIGURE 8.1   Types of machine learning methods applied in human psychology analysis and cognitive science.

which we can detect stress, making it possible to diagnose at a very early stage by a special area study of artificial intelligence, i.e. machine learning. Mental issues lead to serious problems, sometimes requiring cognitive and motor rehabilitation in turn [1, 2]. We know that humans learn from previous experiences, but machinery can learn from the past data as well, yet react in a much faster manner than the humans if they are trained accordingly—that is what we define as machine learning. There are three types of machine learning methods according to their purpose as shown in Figure 8.1 and they are often used in various clinical applications.

The next section provides a detailed survey of recent research conducted to analyze human psychology and cognitive skills. A comparative analysis along with among them has been presented, which can be a guide for researchers working in this domain to proceed further. Different computer-aided diagnosis techniques applied to measure mental disorders have been discussed in Section 8.3. The working principles of the three types of machine learning methods, i.e. reinforcement learning, supervised, unsupervised models have been described with pictorial representations in Section 8.4. And finally, Section 8.5 concludes and elaborates on several advantages and loopholes found from the state-of-the-art experimental outcomes, which can be overcome in the further studies and applications.

8.2   LITERATURE REVIEW

We know that in machine learning, machines learn and improve from experiences rather than programming. Machine learning can recognize patterns, plus it can offer solutions to some complex problems. It is an important tool that can apply to human psychology, analyzing experimental results. In this section, we present an overview that demonstrates when and how machine learning techniques can benefit from human behavioral theories. Here we can't systematically review the present advantage cognitional modeling process using machine learning. Let us focus on the advantage of the most extensively used machine learning methods and use it to analyze the result collected from human psychology. Machine learning can be useful to approximate collecting data and also help to achieve some objectives that can:

- Specify fresh and new data in developing models
- Focus on forecast at the single-subject layer in developing models

Mental illness is one of the biggest issues confronted by a large population of the world today. People can become violent and wild, or sometimes become silent and desperate due to mental illness. When they are violent, they can harm people around them or the hurt or even kill themselves. In essence, they become a psychological patient, so we need psychological analysis to save people from these circumstances. Here, we can use machine learning in human psychology analysis and cognitive science to prevent psychological disorders. Using the rich past of problem descriptions, experimental methods, and theories developed by cognitive psychology specialists to study the human brain, Ritter et al. [3] deal with the interpretability issue in today's deep neural networks (DNNs). Working with the collection of data of stimuli encouraged by the original cognitive psychology research, it was discovered that the state-of-the-art one-shot learning models trained on an ImageNet displayed a similar bias to what is present in human intellect: categorizing different objects by their shapes but not color is what they prefer. Dwyer et al. in [4] have addressed the problems of clinical translational psychiatry and psychology, which can optimally point out with machine learning falling into four major sections: diagnosis, prognosis, cure prediction, discovery and observing the probable biomarkers. Here, the final target of translational machine learning is benefitting the general practitioners, clients, and in specialized hospital settings by invention of some process, for improvement of treatment results. Consider a real-life scenario where to suggest a diagnosis, a decision support aide could use biological or clinical unique identifiers, upcoming prognosis, optimal cures to carry out biological signature observation as an objective surrogate of treatment success. The predictions could be in the form of regression interfaces for conveying uninterrupted approximations (e.g., the patient will get an advantage from a particular dose of medicine X) or classifications (e.g., the person will get an advantage from treatment X). These types of forecasts could also be improved with machine learning practices to identify subsets of individuals (e.g., clustering) or to key individuals against a population norm.

Orru et al. [5] claimed that with the study of machine learning, complementing the analytic workflow of psychological trials both minimize copying ability and maximize precision problems. If compared with statistical deduction, machine learning analysis of experimental data is model agnostic and rather than inference, it is mainly based on prediction. Agrawal et al. outlined a data-driven, iterative method [6] which allows cognitive scientists using machine learning to produce models that are precise and interpretable both, in case of demonstration in the region of moral decision-making, where appropriate principles that manipulate human judgments are often identified by standard experiments, yet, generalization of these discoveries has not been possible in real world scenarios which come up with conflicting principles. A significant step toward adaptive and customized learning is identifying the cognitive profile of a learner. Electroencephalograms

(EEG) is exercised in [7] to identify emotional and cognitive states of a subject. By the use of EEG signals, an advance for detection of two cognitive skills, focused concentration and operational memory, is planned. The classification precisions obtained on 86 subjects were 81% and 84% in case of working memory and focused attention, respectively.

The usage of machine learning algorithms in the domain of psychology is expected to amplify sharply in the near future. Progressive machine learning applications will need association with data scientists, hence, it will be a necessity for the researchers in the domain of character psychology to ready themselves with both the methodology and terminology of machine learning [8]. In the field of human-computer interaction, researchers have basic interest to interact with and use customized systems, where at the same time, cognitive science partially provides the theoretical aspects of different observations [9]. Dupoux et al. [10] analyzed the scenarios where "reverse engineering" language growth, i.e., creation of a scheme, which is capable of mimicking different achievements of infants, has contributed to our logical understanding of quicker language learning. They claim that, from the computational perspective, moving on from toy problems to the full complexity of the learning scenario, and taking as input as correct reconstructions of the sensory signals available to infants as possible, will be better. On the data side, privacy-preserving sources of home data have to be arranged. On the psycholinguistic side, exact tests have to be created to benchmark humans and machines at separate linguistic levels.

In addition to having a better understanding of human mind and mood, the final aim of research in the domain of cognitive science is to eventually develop human-level machine intelligence [11]. At this level, someone would not be able to distinguish the intelligent agent's behavior from human intelligence, an experiment called the Turing test. Emmery et al. [12] in the current commentary on "robust modelling in cognitive science," highlighted the observations where the overlap has occurred and discuss the way of producing novel ongoing challenges by similar proposals, including social change toward open science, the scalability and interpretability of needed practices, and the downstream impact of having robust practices that are totally transparent. Garcia-Ceja et al. [13] have reviewed state-of-the-art research on mental state monitoring with a primary focus on those that use sensors to gather behavioral data and machine learning to analyze these data. It does not include an exhaustive review for each of the specific cases, but presents a subset of relevant works for several of the cases in the context of mental health monitoring applications. The four main application domains that emerged in the literature include (i) detection and diagnosis; (ii) prognosis, treatment and support; (iii) public health, and; (iv) research and clinical administration [14]. The most common mental health conditions addressed included depression, schizophrenia, and Alzheimer's disease.

Jaques et al. [15] have shown that the performance of mood predictor systems using machine learning can be enhanced by customization following some principled processes. Domain adaptation and multitasking learning by a deep learning model can be applied to achieve this customization according to each and every subject's unique mood. The prediction error can be minimized in this way more than the conventional methods, and these personalized models better fit the concerned data by improving the interclass correlation coefficient. This can be a remarkable contribution in healthcare systems as it helps to distinguish stressed and unhappy people so timely treatment can be provided. Some simple modifications in the model future can allow prediction without a user's self-report. The Gaussian process can also be extended to obtain the benefits of multitask learning in the customization process. Table 8.1 shows comparative analysis among some other existing methods in this domain.

TABLE 8.1
Comparative Analysis Between the Existing Methods

Sl. No.	Objective	Method Used	Performance/ Experimental Outcome	Drawback/ Loopholes Found
1. [16]	Mood detection from neurophysical data	Feedforward neural network, convolutional neural network, recurrent neural network (RNN), Long short-term memory along with random forest, SVM, and decision tree	Connection between physical and neurophysical parameters have been confirmed	1 year is a long time for a subject to keep engaged with, GPU can be used in future instead of CPU for larger dataset
2. [17]	Prediction of short-term mood for the next day in a depression context	Data gathered from smartphone and predicting future mood by time series, dynamic time warping, and machine learning classifier	SVM and RF performed better than time series and naïve benchmark	—
3. [18]	Mood prediction from self-reported smart phone and wearable	Classification as high mood or low mood using linear lasso and nonlinear neural network	Existing linear models have been extended by nonlinear Vanilla and RNN	Missing and small dataset. Introduction of multitask learning in more recent data is suggested in future
4. [19]	Prediction of absent mind for mind wandering using EEG	Task-independent electrophysiological markers capable of differentiating mind-wandering from the on-task state	Multifeature SVM has achieved highest accuracy and performed better than the logistic regression classifiers	Introduction of scale (e.g. -5 to 5) to rate participant's concentration power would be more helpful than the present option choosing approach

(Continued)

TABLE 8.1 (*Continued*)
Comparative Analysis Between the Existing Methods

Sl. No.	Objective	Method Used	Performance/ Experimental Outcome	Drawback/ Loopholes Found
5. [20]	Analysis of mental state by delivering questionnaire to a target group of many school, college students and working professionals	Naïve Bayes, SVM, KNN, decision tree, and logistic regression	Random forest, KNN, and SVM almost worked same where application of ensemble classifiers helped to improve the accuracy of mental health prediction	People with more than one mental illness will belong to more than one class, which results in overlap and can be overcome by deep learning or fuzzy in future
6. [21]	Detection of behavioral intentions based on societal influence	Agent-based modelling and simulation (ABMS) and three-level architecture involving system, parts and containing society	Multitiered modelling of social system kept data and privacy constraints outlined	Introduction of PMFserv framework in future will be more effective in assessing the effect of interventions in mental health
7. [22]	Prediction of mental health problem in children	Averaged one-dependence estimator (AODE), multilayer perceptron, radial basis function network, instance-based classifier, K*, multiclass classifier, functional tree and logical analysis data tree	Logical analysis of data tree, multiclass classifier and multilayer perceptron gave more accurate results than the others	The dataset size should be increased for better accuracy in future

8.3 APPLICATION AREAS

People are born with the traces of depression and when this stage is prolonged, people become depressed. To overcome this problem, we need psychological analysis in a regular basis. The application area of machine learning methods in psychology analysis and cognitive science is a wide one and incoudes various aspects. Color psychology is analyzed based on mood [23–25], the mental arithmetic has been used to assess someone's cognitive ability [26–29] and the use of speech and music therapies [30–32, 33], and the impact of mental condition on handwriting has been analyzed using machine learning techniques [34–38]. In addition, sentiment analysis using social media posts [32] and emotion recognition from facial expression are also some common research areas where detection of mood and mental state takes place.

8.3.1 Color Psychology Analysis

A machine learning approach to measure particularity and associative consistency between several colors and emotions has been introduced, and in [23], black and red have shown stronger associations with emotion. Red has an association with both positive and negative emotions while black has shown mostly negative emotions like anger, anxiety, sadness, hatred, etc. Optimized state vector machine (SVM) has been used as the classifier there which can be outperformed further by deep learning method. Visual color perception has been analyzed using EEG spectral features in [24, 25].

8.3.2 Judgment Analysis through Mental Arithmetic

EEG has been used in analyzing and classifying the cognitive load of subjects involved in logical programming in [26]. EEG signals gathered from different subjects have been fuzzified at first and then the resultant fuzzy membership values are passed on as a fuzzy rule-based classifier's input. The approach outperforms SVM and Naïve Bayes as a classifier while decision-making capability and judgment analysis have been performed during coding and debugging of a software developer. In the future, the type-2 fuzzy can be introduced to address intrapersonal uncertainty and feature selection can be done to have more appropriate features in the workspace. A dataset consisting of mathematical problems including arithmetic, algebra, calculus, manipulating polynomial, measurement, comparison, and probability as its components has been introduced in [27] which is extendable due to its modularity. The main aim of Saxton et al., i.e. identifying mathematical reasoning ability of neural models, is achieved here, but linguistic complexity and variation have not been addressed and in the future, visual reasoning can also be taken into account as it is useful in mathematical reasoning problems where a visual format is not specified. Mandal et al. [28] reviewed different mathematical word problems of elementary school level and found that in the earlier stage, the datasets were small and rule-based artificial intelligence was applied in most of the cases. Text classification of machine learning and natural language processing slowly replaced the rule-based process. Different e-learning applications came into the picture in the later period, which has been very effective in our daily life now [29].

8.3.3 Handwriting Analysis

According to psychologists, handwriting plays a vital role in analyzing human nature, personality, and mood. Handwriting strokes with shorter height and width show a negative mood as compared to those made in neutral or positive moods. Additionally, humans write in a slower manner while in a negative frame of mind compared to those in neutral times. Different mental workload scenarios analyzed pressure, spatial, and temporal measures while drawing a figure and at first handwriting a paragraph and then copying that into a digitizer in [34]. The sample size is comparatively less here, which can be increased in future enhancements, and the use of more heterogenous sampling in terms of education degree, age, gender etc. is suggested. There is a rising community of researchers who are focusing on the science of experiential or the lived aspect of the human mind involving brain computer interface as it has a severe impact in the recent methodological and conceptual transition in the cognitive science research domain [35]. Graphology is the domain where handwriting is analyzed by different image processing algorithms like polygonalization, template matching, thresholding method, etc. Based on several factors like letter slant, baseline, margin, "t" bar characteristics, feature extraction, and personality trait mapping using machine learning have been done in [36], which can be useful as automated handwriting analysis tools in the future. The writer has been identified by analyzing and writing, and personality has been attempted to be identified in [37] as handwriting reflects the subconscious mind's expression and can predict human behavior also. Use of artificial neural networks (ANNs) in the future can be an effective way to identify someone's fright, honesty, etc. Convolutional neural networks have been applied to train handwriting dataset in [38] for predicting behavior of different subjects automatically from their handwriting, which reduces the cost of manually analyzing handwriting. However, more features like cursive handwriting can be introduced in future works and the system can be made user-specific to address a wider range of people in this regard.

8.3.4 CONCENTRATION ANALYSIS

Concentration power during performing some task can be analyzed using machine learning tasks to identify someone's mental state. The impact of cognitive load while driving a car has been analyzed and monitored in [39] to avoid road accidents due to alcohol intake, absence of mind, sleepiness, distraction in form of phone use, etc. Using a fuzzy [40] interface system, the driver has been tested in three roadway scenarios, the first one is the baseline, the second one involves visual distraction, and the third one is balanced by a pure cognitive load. The proposed scheme can be an effective one for warning the driver of risky driving and motivating him/her to improve future behavior [32, 41–43].

8.4 MACHINE LEARNING MODELS

The model of machine learning is an integral part of the machine learning method. With the assistance of this model, we can analyze the experimental results that can apply to the human psychological condition. In this section, we explore some steps and guidelines for selecting a machine learning model. The overall steps are:

1. Collect data
2. Check for exceptions, mine data, and clean data
3. Carry out the statistical analysis and initial imagination
4. Choose the model
5. Check the validity
6. Find the actual result

As previously discussed, we can categorize machine learning models used in the concerned domain into three types, i.e. supervised models, unsupervised models, and reinforcement models.

8.4.1 Unsupervised Model

The unsupervised model means there is no supervision i.e. no training is given to the machine and it acts on data that is not labeled [44]. Hence, the machine tries to identify the pattern and then respond. Unsupervised learning is one of many methods by which we can train an artificial intelligence algorithm; it works with unclassified and not labeled information and is then allowed to work with no instruction on that information. Unsupervised models group based on some resemblance (e.g., clustering, anomaly detectors, etc.). The most popular method of clustering is k-means clustering, which originated from signal processing and has been used in mood detection and cognitive ability analysis.

8.4.2 Supervised Model

A supervised model means work is done under supervision i.e. the machine is trained with data that is valuable and then forecasts with the help of a labeled dataset (data for which the target answer is already known as a labeled dataset). Further, supervised models are divided into two subcategories:

a. **Regressor:** When the target variable is a continuous or real value, the encountered problem is a regression problem. Such as monthly income based on work experience.
b. **Classifier:** When the target variable is categorical, the problem is a classification problem. For example, red/blue, yes/no, etc.

Different classification methods used to sense the mood in existing literature are described next.

8.4.2.1 Decision Tree

Decision trees are very popular models that are used not only in machine learning, but also in other fields like strategic planning, operation research, data mining, etc. [45]. Decision trees can be used in the concept of classification and also for regression in the form of a tree structure. Each square of the decision tree is called a node, and the more nodes in a decision tree, the more accurate it will be. The node from which two or more branches emerges is called a decision node because they have to decide "yes" or "no," meaning the machine must decide whether to go to the left or right side of the decision tree. The last nodes, where a conclusion is reached or decision is made, are called the leaf nodes, which stand for a decision or a classification. The tree's highest node is the root node. During the rapid growth of the decision tree, it simultaneously divides a dataset in eventually smaller subsets as shown in Figure 8.2.

8.4.2.2 Naïve Bayes

Naïve Bayes comes under a supervised learning algorithm, which is used for classification purposes [46]. Hence, it is also called as Naïve Bayes classifier. With the help of certain characteristics, it predicts the target variable, just like other supervised learning algorithms. Given a dataset of input features, it finds the observation class for mood and emotion classification. It has been used in different applications.

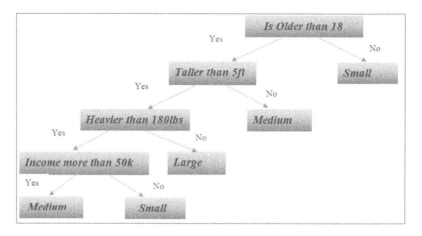

FIGURE 8.2   Pictorial representation of a decision tree.

8.4.2.3   Artificial Neural Network (ANN)

In the short form, an ANNs is an instance of processing information that is influenced through information related to processes in the biological nervous system [47], like our brain. Consisting of massive interconnected processing elements (neurons) working together to solve a specific problem, ANNs were basically developed for the development and testing of computational analogs of neurons by psychologists and neurobiologists. In the phase of learning, ANNs learn by adjusting the weight so that it predicts the exact class label of the input. For regression and classification, ANN can be used.

8.4.2.4   Support Vector Machine (SVM)

SVM takes place under supervised machine learning algorithm [48] as shown in Figure 8.3. It can be used for both classification or regression, but mostly for classification in machine learning. The purpose of the support vector machine algorithm is to find a hyperplane in an n-dimensional space (n = number of properties) that explicitly classifies data points. To distinguish the data point in two classes, there are many possible hyperplanes that can be used. The main goal of SVM is to find a plane that has the highest margin, i.e. the maximum distance between data points of both classes.

8.4.2.5   Random Forest

Random forest belongs to the supervised learning algorithm model, used for both classification and regression in machine learning [49]. Generally, it is used to solve classification problems like SVM. It is based on the concept of ensemble learning. It combines multiple classifiers to solve complex problems and increase the functionality of the model. It contains multiple decision trees in different subsets of a given dataset and takes an average to enhance the predictive accuracy of that dataset. It does not work on only one decision tree, it takes the prediction from every tree and selects the best solution by majority votes of prediction as

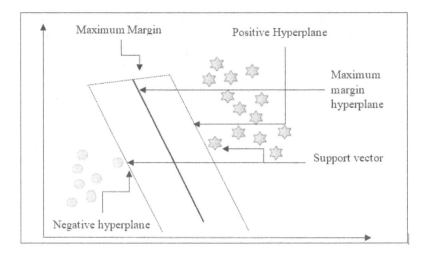

FIGURE 8.3 Support vector machine (SVM) working diagram.

the final output. The higher number of trees get a more accurate result in random forest use.

8.5 CONCLUSION

This chapter has abridged the survey on human behavior and machine learning that provides an exploratory understanding of some important points related to the survey on the effectivity of machine intelligence in mental health improvement. While undertaking the survey, some research challenges have been identified by us and based on that, some future directions can be highlighted as follows:

- Machine intelligence and natural intelligence of humans are different from each other in various aspects like: social intelligence, evolutionary history, consciousness, etc.
- We need ways to evaluate to grow transparent, smart, and easy-to-use benchmarking policies.
- With respect to human vs. machine intelligence, we should move from a competition to an interaction paradigm with collaboration and synergy exploitation between both intelligences, since machine intelligence is in fact a product of human intelligence.
- Further research on how the human-machine interaction impacts human intelligence and cognitive abilities can be undertaken. Additionally research on how artificial intelligence (AI) modifes human relationships and the effect of AI on humans and surroundings is needed. While most of the recent surveys concentrate on the interaction between adults and AI systems, the way children handle artificial systems should be further researched, as it is most important for the next generation.

Finally, if the highlighted points are applied in several areas, we should conclude that some algorithmic aspects are there to be analyzed irrespective of the application background. Multidisciplinary thinking, several different teams, and future impact evaluation are needed for this, since the forthcoming disruption is on its way and we want to see ourselves and our circumstances in a better position.

REFERENCES

1. Ghosh, A. and Saha, S., 2020. Interactive game-based motor rehabilitation using hybrid sensor architecture. In: *Handbook of Research on Emerging Trends and Applications of Machine Learning* (pp. 312–337). IGI Global.
2. Saha, S. and Ghosh, A., 2019. Rehabilitation using neighbor-cluster based matching inducing artificial bee colony optimization. In: *2019 IEEE 16th India Council International Conference (INDICON)* (pp. 1–4). IEEE.
3. Ritter, S., Barrett, D.G., Santoro, A. and Botvinick, M.M., 2017. Cognitive psychology for deep neural networks: A shape bias case study. In: *Proceedings of the 34th International Conference on Machine Learning, Volume 70* (pp. 2940–2949). JMLR.org.
4. Dwyer, D.B., Falkai, P. and Koutsouleris, N., 2018. Machine learning approaches for clinical psychology and psychiatry. *Annual Review of Clinical Psychology, 14*, pp. 91–118.
5. Orrù, G., Monaro, M., Conversano, C., Gemignani, A. and Sartori, G., 2020. Machine learning in psychometrics and psychological research. *Frontiers in Psychology, 10*, p. 2970
6. Agrawal, M., Peterson, J.C. and Griffiths, T.L., 2019. Using machine learning to guide cognitive modeling: A case study in moral reasoning. *arXiv preprint arXiv: 1902.06744.*
7. Mohamed, Z., El Halaby, M., Said, T., Shawky, D. and Badawi, A., 2018. Characterizing focused attention and working memory using EEG. *Sensors, 18*(11), p. 3743.
8. Stachl, C., Pargent, F., Hilbert, S., Harari, G.M., Schoedel, R., Vaid, S., Gosling, S.D. and Bühner, M., 2019. Personality research and assessment in the era of machine learning. *European Journal of Personality, 34*(5), p. 613–631.
9. Zanker, M., Rook, L. and Jannach, D., 2019. Measuring the impact of online personalisation: Past, present and future. *International Journal of Human-Computer Studies, 131*, pp. 160–168.
10. Dupoux, E., 2018. Cognitive science in the era of artificial intelligence: A roadmap for reverse-engineering the infant language-learner. *Cognition, 173*, pp. 43–59.
11. Beriha, S.S., 2018. Computer aided diagnosis system to distinguish ADHD from similar behavioral disorders. *Biomedical & Pharmacology Journal, 11*(2), p. 1135.
12. Emmery, C., Kádár, Á., Wiltshire, T.J. and Hendrickson, A.T., 2019. Towards replication in computational cognitive modeling: A machine learning perspective. *Computational Brain & Behavior, 2*(3-4), pp. 242–246.
13. Garcia-Ceja, E., Riegler, M., Nordgreen, T., Jakobsen, P., Oedegaard, K.J. and Tørresen, J., 2018. Mental health monitoring with multimodal sensing and machine learning: A survey. *Pervasive and Mobile Computing, 51*, pp. 1–26.
14. Shatte, A.B., Hutchinson, D.M. and Teague, S.J., 2019. Machine learning in mental health: A scoping review of methods and applications. *Psychological Medicine, 49*(9), pp. 1426–1448.
15. Jaques, N., Taylor, S., Sano, A. and Picard, R., 2017. Predicting tomorrow's mood, health, and stress level using personalized multitask learning and domain adaptation. In: *Proceedings of IJCAI 2017 Workshop on Artificial Intelligence in Affective Computing, Proceedings of Machine Learning Resarch* (pp. 17–33).

16. Kilimci, Z.H., Güven, A., Uysal, M. and Akyokus, S., 2019. Mood detection from physical and neurophysical data using deep learning models. *Complexity, 2019.*
17. van Breda, W., Pastor, J., Hoogendoorn, M., Ruwaard, J., Asselbergs, J. and Riper, H., 2016. Exploring and comparing machine learning approaches for predicting mood over time. In: *International Conference on Innovation in Medicine and Healthcare* (pp. 37–47). Springer, Cham.
18. Shi, M., 2019. Machine learning methods for mood prediction using data from smartphones and wearables. Thesis, University of Michigan, Ann Arbor.
19. Jin, C.Y., Borst, J.P. and van Vugt, M.K., 2019. Predicting task-general mind-wandering with EEG. *Cognitive, Affective, & Behavioral Neuroscience, 19*(4), pp. 1059–1073.
20. Srividya, M., Mohanavalli, S. and Bhalaji, N., 2018. Behavioral modeling for mental health using machine learning algorithms. *Journal of Medical Systems, 42*(5).
21. Luxton, D.D., 2016. An introduction to artificial intelligence in behavioral and mental health care. In: *Artificial Intelligence in Behavioral and Mental Health Care* (pp. 1–26). Academic Press.
22. Sumathi M.R., Poorna B., 2016. Prediction of mental health problems among children using machine learning techniques. *International Journal of Advanced Computer Science and Applications, 7*(1).
23. Jonauskaite, D., Wicker, J., Mohr, C., Dael, N., Havelka, J., Papadatou-Pastou, M., Zhang, M. and Oberfeld, D., 2019. A machine learning approach to quantify the specificity of colour–emotion associations and their cultural differences. *Royal Society Open Science, 6*(9), p. 190741.
24. Paulraj, M.P., Adom, A.H., Hema, C.R. and Purushothaman, D., 2012. Analysis of Visual Colour Perception using EEG Spectral Features. *Karpagam Journal of Computer Science, 6*(2), pp. 74–81.
25. Rafegas, I. and Vanrell, M., 2018. Color encoding in biologically-inspired convolutional neural networks. *Vision Research, 151*, pp. 7–17.
26. Chatterjee, D., Sinharay, A. and Konar, A., 2013, July. EEG-based fuzzy cognitive load classification during logical analysis of program segments. In: *2013 IEEE International Conference on Fuzzy Systems (FUZZ-IEEE)* (pp. 1–6). IEEE.
27. Saxton, D., Grefenstette, E., Hill, F. and Kohli, P., 2019. Analysing mathematical reasoning abilities of neural models. *arXiv preprint arXiv:1904.01557.*
28. Mandal, S. and Naskar, S.K., 2019. Solving arithmetic mathematical word problems: A review and recent advancements. In: *Information Technology and Applied Mathematics* (pp. 95–114). Springer, Singapore.
29. Wang, Y., Widrow, B., Zadeh, L.A., Howard, N., Wood, S., Bhavsar, V.C., Budin, G., Chan, C., Fiorini, R.A., Gavrilova, M.L. and Shell, D.F., 2016. Cognitive intelligence: Deep learning, thinking, and reasoning by brain-inspired systems. *International Journal of Cognitive Informatics and Natural Intelligence, 10*(4), pp. 1–20.
30. Janvale, G.B., Gawali, B.W., Deore, R.S., Mehrotra, S.C., Deshmukh, S.N. and Marwale, A.V., 2010. Songs induced mood recognition system using EEG signals. *Annals of Neurosciences, 17*(2), p. 80.
31. Hou, Y. and Chen, S., 2019. Distinguishing different emotions evoked by music via electroencephalographic signals. *Computational Intelligence and Neuroscience, 2019.*
32. Biswas, S., Ghosh, A., Chakraborty, S., Roy, S. and Bose, R., 2020. Scope of sentiment analysis on news articles regarding stock market and GDP in struggling economic condition. *International Journal, 8*(7).
33. Lin, C., Liu, M., Hsiung, W. and Jhang, J., 2016. Music emotion recognition based on two-level support vector classification. In: *2016 International Conference on Machine Learning and Cybernetics (ICMLC)*, 1, pp. 375–389. IEEE.
34. Rosenblum, S. and Luria, G., 2016. Applying a handwriting measurement model for capturing cognitive load implications through complex figure drawing. *Cognitive Computation, 8*(1), pp. 69–77.

35. Froese, T., Suzuki, K., Ogai, Y. and Ikegami, T., 2012. Using human-computer interfaces to investigate 'mind-as-it-could-be' from the first-person perspective. *Cognitive Computation*, *4*(3), pp. 365–382.

36. Joshi, P., Agarwal, A., Dhavale, A., Suryavanshi, R. and Kodolikar, S., 2015. Handwriting analysis for detection of personality traits using machine learning approach. *International Journal of Computer Applications*, 130(15).

37. Patil, V. and Mathur, H., 2020. A Survey: Machine Learning Approach for Personality Analysis and Writer Identification through Handwriting. In *2020 International Conference on Inventive Computation Technologies (ICICT)* (pp. 1–5). IEEE.

38. Chanchlani A., Jaitly A., Kharade P., Kapase R., Janvalkar S., 2018. Predicting human behavior through handwriting. *International Journal for Research in Applied Science & Engineering Technology*, 6(6), pp. 849–854

39. Sena, P., d'Amore, M., Pappalardo, M., Pellegrino, A., Fiorentino, A. and Villecco, F., 2013. Studying the influence of cognitive load on driver's performances by a Fuzzy analysis of Lane Keeping in a drive simulation. *IFAC Proceedings Volumes*, *46*(21), pp. 151–156.

40. Ghosh, A., Saha, S. and Konar, A., 2020. Fuzzy posture matching for pain recovery using yoga. In: *Computational Intelligence in Pattern Recognition* (pp. 957–967). Springer, Singapore.

41. Palinko, O., Kun, A.L., Shyrokov, A. and Heeman, P., 2010. Estimating cognitive load using remote eye tracking in a driving simulator. In: *Proceedings of the 2010 Symposium on Eye-tracking Research & Applications* (pp. 141–144).

42. Lan, T., Adami, A., Erdogmus, D. and Pavel, M., 2005. Estimating cognitive state using EEG signals. In: *2005 13th European Signal Processing Conference* (pp. 1–4). IEEE.

43. Cernian, A., Olteanu, A., Carstoiu, D. and Mares, C., 2017. Mood detector-on using machine learning to identify moods and emotions. *In 2017 21st International Conference on Control Systems and Computer Science (CSCS)* (pp. 213–216). IEEE.

44. Khanum, M., Mahboob, T., Imtiaz, W., Ghafoor, H.A. and Sehar, R., 2015. A survey on unsupervised machine learning algorithms for automation, classification and maintenance. *International Journal of Computer Applications*, 119(13).

45. Sharma, H. and Kumar, S., 2016. A survey on decision tree algorithms of classification in data mining. *International Journal of Science and Research*, *5*(4), pp. 2094–2097.

46. Rish, I., 2001. An empirical study of the naive Bayes classifier. *In: IJCAI 2001 Workshop on Empirical Methods in Artificial Intelligence, 3*(22), pp. 41–46.

47. Zhang, G.P., 2000. Neural networks for classification: A survey. *IEEE Transactions on Systems, Man, and Cybernetics, Part C (Applications and Reviews)*, *30*(4), pp. 451–462.

48. Yang, Y., Li, J. and Yang, Y., 2015. The research of the fast SVM classifier method. In: *2015 12th International Computer Conference on Wavelet Active Media Technology and Information Processing (ICCWAMTIP)* (pp. 121–124). IEEE.

49. Ali, J., Khan, R., Ahmad, N. and Maqsood, I., 2012. Random forests and decision trees. *International Journal of Computer Science Issues*, *9*(5), p. 272.

9 Speech Emotion Recognition using Manta Ray Foraging Optimization Based Feature Selection

Soham Chattopadhyay, Arijit Dey, Hritam Basak, and Sriparna Saha

CONTENTS

9.1 INTRODUCTION

Expressing emotions is the most fundamental part of conveying messages among the human and nonhuman primates and it is the most common way to express love, sorrow, anger, hatred, or other states of mind [1, 2]. Even nonspeaking living beings also find ways to express their emotions. We find verbal communication and associated emotions so important that we often miss the scarcity of those in text messages or emails and hence switch to the usage of emojis. With the recent advancements in natural language processing and speech to text conversions, scientists have often looked upon the automation of information parsing from speech audio and artificial intelligence (AI) has been successfully deployed for generating auto-replies; the chatbots and recent speaking-humanoid robots are exemplary evidence of these advancements. However, the emotion analysis of auditory signals has been studied lately by researchers and several improvements have been made in this domain for the last two decades.

Although Speech Emotion Recognition (SER) has versatile applications [3, 4, 5, 6], there exists no generalized or common consensus on categorization and classification of emotions from speech signals as the emotions are a subjective property of humans. Emotions may vary in intensity, mode of expression from person to person, and may vary or are frequently misinterpreted by people. Therefore, the automatic AI-based classification of emotion from auditory signals is the test bed of assessing the performance of several existing feature extraction and classification methods [7, 8]. Though the discrete and dimensional models have shown improved performance in recent times, there is a huge scope of improvements as it remains an open problem in a bird's-eye view.

The human voice can have features from different modalities. The most predominant ones are (1) voice quality, (2) Teager energy operator, (3) prosodic, and (4) spectral features [9]. The robustness and less redundancy of features help to improve classification accuracy. Supervised learning is based on the feature quality and accurately labeled dataset and the performance of these classification problems has highly relied upon the efficacy and experience of the feature-engineer performing the feature extraction as more imagination can open up the possibility of new multimodal features extraction. But this requires lots of time and imagination and still, it is impossible to extract all the important, high-quality, and insightful features from auditory signals.

9.1.1 CONTRIBUTION OF THIS CHAPTER

The contribution of this chapter can, therefore, be considered as a twofold contribution: First, we explore the feature concatenation in auditory signal processing, which to our knowledge, has not been investigated intensively before this work. Merging of features from different sources has been successfully utilized in the image processing task before [1, 2, 10, 11, 12]. Being inspired by these, we have successfully evolved our proposed approach based on feature concatenation. Secondly, we have proposed a bio-inspired meta-heuristic Manta Ray Foraging Optimization (MRFO) algorithm for feature selection and removal of redundant features. This algorithm was evolved in 2020 by Zhao et al. [13], being inspired by the behavioral study of an aquatic species named Manta Ray.

Our proposed model outperformed most of the existing techniques in speech emotion analysis and offers a state-of-the-art result in the publicly available SAVEE and EmoDB datasets. We achieved a classification accuracy of 97.49% and 97.68% on SAVEE and EmoDB datasets, respectively.

9.2 LITERATURE SURVEY

The research in this field started in the early 1990s, whereas the first significant result was obtained by Nakatsu et al. [14] in 1999 where they had used speech power and basic Linear Predictive Coding (LPC) features from 100 human specimens, equally distributed among males and females. The simplest neural network model they used, produced a result of 50% recognition accuracy. Later, Schuller et al. [15] proposed a hidden-Markov model, and validation was done on the EmoDB and VAM datasets that produced correct classification accuracy of 76.1% and 71.8%, respectively, using raw contours of zero crossing rate (ZCR), energy, and pitch features. Later, the improved version of their proposed Markov model produced a state-of-the-art result on German and English speech signals containing 5,250 samples with an average accuracy of 86.8% with global prosodic features using continuous HMM classifier and pitch and energy-based features [16–20]. Rong et al. [3] had used KNN classifiers for classifying the Mandarin dataset using ZCR, spectral, and energy features. Rao et al. [4] used the Support Vector Machine classifier using RBF kernel with approximately 67% classification accuracy using prosodic features. LFPC features were extracted and the HMM classifier was used for the classification of the Mandarin language by Nwe et al. [21] with an average precision of 78.5%. Another important result was produced by Wu et al. [7] by using an SVM classifier that reached a classification accuracy of 91.3% on the EmoDB dataset and 86% accuracy on the VAM dataset by using prosodic, ZCR, and Teager energy operator (TEO) features. Deng et al. [22] validated their proposed method on four different datasets: EmoDB, VAM, AVIC, and SUSAS by using autoencoder classifiers. The features used were MFCC and other LLD features (example: ZCR and pitch frequency etc.) and denoising was performed before the classification task and presented classification accuracy of 58.3%, 60.2%, 63.1%, and 58.9% respectively.

With the recent advancement in deep learning in the last few years, scientists have tried to exploit the ability of deep neural network (DNN) models to learn high-quality semantic information and invariant features from different types of datasets [7, 23–24]. A few recent studies provided results that supported this conclusion that DNNs are equally efficient and useful for speech-emotion classification. Rozgić et al. [25] trained the DNNs using the combination of lexical and acoustic features. Unlike these existing DNN models that directly used acoustic features learned from sources, our proposed method uses optimization in between for the performance improvement.

9.3 MATERIALS AND METHODS

Here, we describe the workflow of this experiment with the following steps: (1) Dataset description (2) preprocessing, (3) feature extraction, (4) feature selection using optimization algorithm, and (5) classification and analysis of results. The schematic representation of the proposed work is in Figure 9.1.

9.3.1 DATASET DESCRIPTION

The initial step of the experiment is dataset collection and preprocessing. We used two publicly available datasets for this experimentation purpose. The datasets are widely used by several other experimentations worldwide before our work. The two datasets used in this experimentation are named SAVEE and Berlin EmoDB (widely known as "EmoDB" dataset). However, we plan for further validation of our method

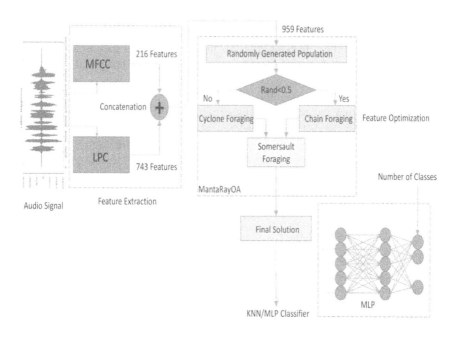

FIGURE 9.1 Workflow of our proposed method.

on other publicly available datasets in the near future. The brief description of these two datasets and the classes of emotion signals are described below.

9.3.1.1 SAVEE Dataset

The Surrey Audio-Visual Expressed Emotion (SAVEE) Database consists of speech recordings from four British actors consisting of 480 samples in total, collected and labeled with extreme precision and by using high-quality equipment. The dataset is classified into seven different emotional categories: happiness, sadness, fear, neutral, disgust, anger, and surprise. The sentences were chosen by experts carefully from Texas Instruments and Massachusetts Institute of Technology (TIMIT) Corpus and were balanced phenotypically in every category.

9.3.1.2 EmoDB Dataset

The Berlin EmoDB dataset contains emotion speech data from 10 different speakers and contains 500 labeled samples in total. It also contains 7 categories of emotion-speeches: normal, anger, sadness, happiness, disgust, anxiety, and fear.

9.3.2 PREPROCESSING

For every kind of signal processing task, the preprocessing of sample data has an important role in determining the performance of a model. This includes denoising, filtering, signal enhancement, amplitude or frequency modulation, and visual representation of signals for better understanding. Later, normalization, framing, and windowing are done for selecting the specific region of interest of the available signal dataset. Some simple audio preprocessing techniques which have improved the exploitation of our model are discussed next.

9.3.2.1 Pre-emphasize

The pre-emphasis step is mainly carried out as it synthesizes the normal form of any amplitude signal. The main idea behind this is to flatten the speech spectrum, which can be done by implementing a high-pass Finite Impulse Response (FIR) filter. The expression of the filter in a discrete frequency domain is given by:

$$F(z) = 1 - Az^{-1} \tag{9.1}$$

To normalize the signal, first the maximum value of the signal has to be taken as the nominator and then divide the signal with it. Thus, the entire signal is normalized between -1 and 1. For smooth transactions between frames, 50% overlaps of consecutive frames are accepted. The mechanical acoustic signal can be stable in the range of 50 ms to 200 ms, so we selected a short window for better feature extraction.

9.3.2.2 Framing

For further processing, the signal is divided into small frames so a sequence of frames forms the entire original signal. This is done so that a long signal can be analyzed independently in small frames and can be expressed through a single feature vector [26–28]. Some aspects of framing, like frameshift, are the time difference between two starting points of two consecutive frames and frame length is the duration of time for each frame.

9.3.2.3 Windowing

For audio signals, it is quite common to have discontinuities at the frame edges of the signal. This phenomenon often causes bad performance in the audio processing task. To remove these discontinuities, trapped windows such as the Hamming window are applied at every frame. The general expression of the Hamming window is:

$$W_h = a - b \times cos\left(\frac{2\pi n}{N-1}\right) \tag{9.2}$$

where $a = 0.54$, $b = 0.46$, and N denotes the frequency of the sample space in a partition of the data in some random complementary subsets.

9.3.3 FEATURE EXTRACTION

Features have an important role in any classification task as all the information of speech data is embedded in these features. Therefore, high-quality and accurately extracted features may contribute immensely toward the improved performance of a classifier. In this experiment, we have extracted two short-time features from the dataset: MFCC and LPC. A brief description of the feature-extraction is described next.

9.3.3.1 Mel Frequency Cepstral Coefficient (MFCC) Features

MFCC features are based on human auditory sensation characteristics. The mathematical simulation of hearing is done in MFCC by using some nonlinear frequency units. We normally use the fast Fourier transform (FFT) or discrete Fourier transform for the conversion of acoustic signals from time-domain data into frequency-domain data for each sample frame. The FFT is given by the following equation:

$$x(j) = \sum_{p=0}^{n-1} x[i] e^{-j\frac{2\pi p j}{n}} \tag{9.3}$$

where the time-domain signal is represented by $x[i]$, which transforms into frequency domain signal $x(j)$, and n is the number of samples in every frame. Next, we calculate the Disperse Power Spectrum by using the following equation:

$$Power\ Spectrum\ \left(PS_p\right) = x(j). \wedge x(j) \tag{9.4}$$

Thereafter, we have the Melody spectrum by the PS_p in the triangular filter bank, which consists series of triangular filters with cutoff frequencies. These frequencies are determined using center frequencies of two consecutive filters. These filters are linear in Melody frequency coordinates. The scale is equivalent to the span of every filter and the value for the span is set to 167.859 in this project. The frequency response of the triangular filter can be calculated as:

$$F[n] = \begin{cases} 0, n < g(p-1) \\ \dfrac{2(n-g(p-1))}{(g(p+1)-g(p-1))\times(g(p)-g(p-1))}, g(p-1) \leq n \\ \dfrac{2(g(p+1)-n)}{(g(p+1)-g(p-1))\times(g(p+1)-g(p))}, g(p) \leq n \\ 0, n > g(p+1) \end{cases} \qquad (9.5)$$

where $p = 1, 2, ..., 12$, $g(p)$ is the center frequency of the filter, $n = 1, 2, ..., (N/2\text{-}1)$ where N is the number of samples per frame.

To improve the quality of the features, we used the logarithmic spectrum of power spectrum on the signal that represents by the following equation.

$$L(p) = \ln\left(\sum_{n=0}^{N-1} |PS_p|^2 F[n]\right), 0 \leq p \leq N \qquad (9.6)$$

where $L(p)$ is the logarithmic spectrum, $F[n]$ is the series of filters, PS_p is defined earlier, N is the number of samples per frame.

Finally, the discrete cosine transform (DCT) of the logarithmic spectrum of the filter banks is calculated that gave the MFCC feature described in the following equation.

$$F[n] = \sum_{p=1}^{N-1} L(p)\cos\left(\dfrac{(M-1)\times k \times \pi}{2N}\right), 0 \leq k \leq N \qquad (9.7)$$

9.3.3.2 LPC Features

The speech signal is sequential data and so we assume the voice acoustics of the nth speech sample is $P[n]$, which can be shown as the combination of previous k samples. The nth speech sample can be mathematically represented as:

$$S[k] = \sum_{j=1}^{M} a_j S[k-j] + h \times E[k] \qquad (9.8)$$

where $j = 1, 2, 3 ... k$, is the jth test sample, h stands for gain factor, $E[k]$ denotes the excitation of the kth sample, and a_j is the vocal tract vector coefficient. LPC is also known as inverse filtering because it determines all zero filters, which are also inverse of the vocal tract model.

9.3.4 MANTA RAY OPTIMIZATION

In this work, we have used the Manta Ray Foraging Optimization (MRFO) algorithm for feature selection. It is a bio-inspired optimization algorithm that works on three unique foraging strategies of manta ray.

9.3.4.1 Chain Foraging

Manta rays have a unique approach toward their prey, which is imitated in our project. They tend to swim toward the highest concentration of planktons, i.e. the best possible position of food. In our algorithm, we set some random initial values and ask them to move toward the optimal solution in every iteration. Though the actual optimal solution is not known, this algorithm assumes that the best solution is the one where the plankton concentration is highest. All the members of the group proceeded toward the plankton concentration, by following the previous member of the group, except the first one. This is known as forming a foraging chain, which is imitated in our experimentation. Mathematically, the chain foraging is represented by:

$$R_i^j(p+1) = \begin{cases} R_i^j(p) + b\left(R_{best}^j(p) - R_i^j(p)\right) + c\left(R_{best}^j(p) - R_i^j(p)\right), \; i = 1 \\ P_j^k(n) + a\left(P_{j-1}^k(n) - P_j^k(n)\right) + c\left(P_{best}^k(n) - P_j^k(n)\right), \; j = 2,3,\ldots,n \end{cases}$$

$$(9.9)$$

$$b = 2a\sqrt{\left|\ln(a)\right|} \qquad (9.10)$$

where $R_i^j(p)$ is the position of an ith agent in j^{th} dimension, b is the random vector in closed range in 0 to 1, c is known as weight coefficient, $P_{best}^k(n)$ is the best plankton concentration position. The location of the $(i+1)^{th}$ is calculated based on previous i agents.

9.3.4.2 Cyclone Foraging

When a group of manta rays discovers a high concentration of plankton in the water, they make a chain-like structure that looks like a cyclone and head toward the plankton. Each of the agents follows the previous agent toward the prey and forms cyclone foraging. Every manta ray doesn't only form the spiral but also follows the same path and moves one step toward the plankton following the one in front of it. The mathematical expression for two-dimensional cyclone foraging is given below.

$$\begin{cases} y_i(p+1) = y_{best} + b\left(y_{i-1}(p) - y_i(p)\right) + \left(y_{best} - y_i(p)\right) e^{cz} \cos(2\pi z) \\ z_i(p+1) = z_{best} + b\left(z_{i-1}(p) - z_i(p)\right) + \left(z_{best} - z_i(p)\right) e^{cz} \cos(2\pi z) \end{cases}$$

$$(9.11)$$

where z is a random value in a closed range between 0 and 1. This kind of spiral foraging algorithm was also explained by [29] for Grey Wolf Optimization, however, this is different.

The N-dimensional form of the equation is as followed:

$$R_i^j(p+1) = \begin{cases} R_{best}^j + b\left(R_{best}^j(p) - R_i^j(p)\right) + \delta\left(R_{best}^j(p) - R_i^j(p)\right), \ p = 1 \\ R_{best}^j + b\left(R_{i-1}^j(p) - R_i^j(p)\right) + \delta\left(R_{best}^j(p) - R_i^j(p)\right), \ p = 2,3,\ldots,n \end{cases}$$

$$\text{(9.12)}$$

$$\gamma = 2e^{c\frac{I-n+1}{I}}\sin(2\pi c) \tag{9.13}$$

In the above equation, δ is the coefficient of weight, max iteration is represented by I, and c is a random parameter having a value in $[0,1]$.

Each of the search agents performs independent exploration between the present position and the position of the target. Therefore, this algorithm can efficiently find the best solution in this range. However, we can force any agent to take a new position that is far from its current position by the following equation:

$$R_r^j = B_{Low}^k + b\left(B_{Upper}^j - B_{Low}^j\right) \tag{9.14}$$

where, R_r^j is the newly specified random position of the jth in the n-dimensional space. B_{Low}^j and B_{Upper}^j are the lower and upper bounds, respectively, of the N-dimensional space. Thus, this algorithm is suitable for finding any best solution in this N-dimensional space.

9.3.4.3 Somersault Foraging

In this type of foraging, the best concentration of plankton is considered as a pivot point. All manta rays move to the point and eventually update their position around the position of the high plankton concentration (i.e. best solution). Somersault foraging is represented by the equation below:

$$R_i^j(p+1) = R_i^j(p) + G\left(dR_{best}^j(p) - gR_i^j(p)\right), \ i = 1, 2, \ldots, n \tag{9.15}$$

where G is known as the Factor of somersault, d and g are random parameters in the closed range between 0 and 1. As the equation suggests, this algorithm allows the search agents to update their position at any possible position in the range of the actual position and the highest concentrated plankton with the reduction in the distance between the agents and the optimal solution. The overall optimization algorithm is given in Figure 9.2.

9.3.5 CLASSIFICATION

In this current work, we have used two different classifiers, one is K-Nearest Neighborhood (KNN) classifier and the other one is Multilayer Perceptron (MLP).

9.3.5.1 KNN Classifier

KNN is a type of supervised machine learning model that finds the "feature similarity" between different data points to predict a new data point with a value based on how closely it matches with the other data points in the training dataset. The KNN algorithm has some inflexible properties like it is very slow to learn. The fact that it has no proper phase of training and classifies with the entire training dataset as a whole, justifies its slow learning very clearly. To implement this algorithm, we have to firstly choose a certain value of K that is the number of nearest neighbor points. Thereafter, using anyone of the Euclidean, Hamming, or Manhattan distance methods, the distance between the test data and each row of training is calculated. Then the points are sorted based on their distance values and top K values are chosen amongst them.

9.3.5.2 MLP Classifier

Multilayer Perceptron (MLP), colloquially known as a "Vanilla Network" is a type of feed-forward artificial neural network. MLP uses back-propagation (a supervised learning technique) to train its weights and biases of different layers. In the recent past, frequently used activation functions of MLP were sigmoid or tanh, but ReLU or Leaky ReLU have proved better in performance and are now used most commonly. However, in the classification layer, sigmoid or softmax activation functions are still used. In this algorithm, changes in each weight in each layer are done by a certain technique called gradient descent.

9.4 EXPERIMENTS

As we propose a novel approach for the emotion recognition task, we had to perform numerous experiments for the fine-tuning and validation of results. As there exists no such prior work for reference, we had to change several parameters in our experiments and observe the changes in model performance in terms of accuracy and other evaluation metrics due to the lack of pre-existing standard parameter values. Some of these experimentations resulted in very poor results and were then discarded from comparison for their redundant nature. The experiments that justify the authenticity of the proposed method are described next.

9.4.1 PARAMETER SELECTION FOR MANTA RAY OPTIMIZATION

The Manta Ray optimization algorithm as shown in Figure 9.2 relies upon the initialized parameters (e.g. omega value, somersault constant, population size) that often decide the convergence time as well as model performance. The omega value can lie between 0 and 1, hence some stochastic values were selected between 0 and 1 and the classification accuracies were measures after feature selection. It was found experimentally that the omega value, when set to 0.75, produced the highest accuracy for both the datasets.

The somersault constant (s) has a value between 1 and 10 and it was found that the optimal point was achieved for the SAVEE dataset when s = 2, whereas when s = 3, the result achieved the highest classification accuracy on the EmoDB dataset.

Initialize the size of the population N, the max iteration T, every manta ray is represented by $p_j(n) = p_1 + rand(p_u . p_1)$ in range j =1,2,3,...,N and n=1. For calculating fitness function each individual $f_i = f(p_i)$ and the best solution till now p_{best} , where p_u and p_1 are the upper bound and the lower bound

WHILE condition is true

    For i=1 to N

        if rand < 0.5 THEN

            if n/T_{max} < rand THEN

$$p_{rand} = p_1 + rand(p_u . p_1)$$

$$p_j(n+1) = \begin{cases} p_{rand}(n) + a.\left(p_{rand}(n) - p_j(n) + b\left(p_{rand}(n) - p_j(n)\right)\right) & j = 1 \\ p_{rand}(n) + a.\left(p_{j-1}(n) - p_j(n) + b\left(p_{rand}(n) - p_j(n)\right)\right) & j = 2,3,..,N \end{cases}$$

        ELSE

$$p_j(n+1) = \begin{cases} p_{best}(n) + a.\left(p_{best}(n) - p_j(n) + b\left(p_{best}(n) - p_j(n)\right)\right) & j = 1 \\ p_{best}(n) + a.\left(p_{j-1}(n) - p_j(n) + b\left(p_{best}(n) - p_j(n)\right)\right) & j = 2,3,..,N \end{cases}$$

            END IF

        ELSE

$$p_j(n+1) = \begin{cases} p_j(n) + a.\left(p_{best}(n) - p_j(n) + b\left(p_{best}(n) - p_j(n)\right)\right) & j = 1 \\ p_j(n) + a.\left(p_{j-1}(n) - p_j(n) + b\left(p_{best}(n) - p_j(n)\right)\right) & j = 2,3,..,N \end{cases}$$

        END IF

    Compute individual fitness function $f(p_i(n+1))$ IF $f(p_i(n+1)) < f(p_{best})$

        THEN $p_{best} = p_i(t+1)$

    FOR j=1 to N

    $P_i(n+1) = p_i(n) + m (k_1 . p_{best} - k_2 . p_i(n))$

    compute individual fitness function $f(p_i(n+1))$ IF $f(p_i(n+1)) < f(p_{best})$

    THEN $p_{best} = p_i(n+1)$

    END FOR

    END WHILE

FIGURE 9.2 Manta Ray optimization algorithm.

Similarly, the population size was set to 30 in both cases for the best classification accuracy. Figure 9.3 [1] describes the results achieved by varying the parameters of the Manta Ray optimizer.

9.4.2 CLASSIFIER HYPERPARAMETER SELECTION

The selection of hyperparameters is an important tuning before classification. In our proposed method, we have used MLP and KNN classifiers for the classification task of signals and hence quite a few experimentations were performed before selecting the optimal one. Figure 9.3 [2] is a comprehensive graphical analysis of the performance of our method with a varying number of hidden layers of MLP and the number of nearest neighbors of the KNN classifier. This is also justified by

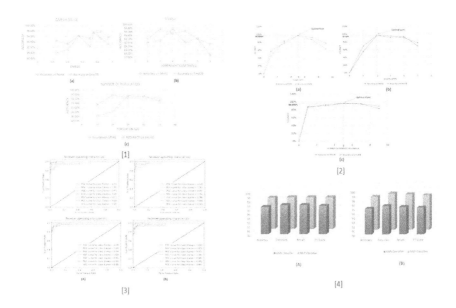

FIGURE 9.3 **[1]** Hyperparameter selection for Manta Ray optimizer: graphical representation of (a) classification accuracy vs. omega value (b) classification accuracy vs. somersault constant(s), and (c) classification accuracy vs. population size. **[2]** Graphical representation of (a) classification accuracy vs. hidden units in each layer of MLP (b) classification accuracy vs. the number of hidden layers of MLP, and (c) classification accuracy vs. the number of nearest neighbors of KNN classifier. **[3]** Receiver operating characteristics of KNN classifier on (a) EmoDB dataset and (b) SAVEE dataset: receiver operating characteristics of MLP classifier on (c) EmoDB dataset and (d) SAVEE dataset. **[4]** Comparative study of the performance of classifiers (i.e. MLP and KNN) with the same features on two different datasets (a) EmoDB dataset and (b) SAVEE dataset.

Table 9.1, which depicts the optimal features selected for classification are (a) hidden units of each layer in MLP = 5, (b) number of hidden layers of MLP = 2, and (c) the number of nearest neighbors of KNN classifier = 5. The parameter selection was done accordingly that produced the best results. The bold digits in Table 9.1 represent the selected hyperparameters and the corresponding classification accuracies.

The classification of different audio signals is mainly done by the feed-forward neural classifier, specifically the MLP. As discussed, our complete work is divided into some main parts. The hyperparameters for MLP and KNN are given in Table 9.2.

9.4.3 FEATURE EXTRACTION

For the feature extraction part, we have particularly extracted MFCC and LPC features separately and concatenated them to get the final feature set. From MFCC and LPC methods we got 216 and 743 features respectively. Therefore, after concatenation, we had total features of 959. Usually, around 13 features are extracted as tabulated in Table 9.2 using the LPC feature extraction methods as we are implementing

TABLE 9.1

Variation of Different Evaluation Parameters with Varying Hyperparameters. The bold numbers are the optimal results achieved.

Parameters	Parameter	SAVEE dataset				EmoDB dataset			
		Accuracy:	Precision:	Recall:	F1 Score:	Accuracy:	Precision:	Recall:	F1 Score:
Number of nearest neighbors of KNN classifier	1	85.63%	86	86	85	87.75%	88	88	88
	3	92.61%	93	93	93	90.43%	91	90	90
	5	**96.55%**	97	97	98	**96.06%**	97	97	97
	7	95.43%	96	96	96	94.91%	95	95	95
	9	89.21%	90	90	90	83.52%	84	83	94
Number of hidden layers in MLP classifier	1	55.03%	56	56	56	69.35%	70	70	69
	2	**97.49%**	98	98	99	**97.68%**	99	98	98
	3	91.35%	91	92	92	95.73%	96	95	96
	4	93.43%	94	94	95	90.53%	90	91	91
	5	69.21%	70	70	70	77.52%	78	78	78
Number of hidden units in each layer of MLP classifier	1	45.03%	45	45	46	57.35%	58	58	58
	3	82.91%	83	83	82	78.35%	79	79	78
	5	**97.49%**	98	98	98	**97.68%**	98	99	98
	7	87.43%	88	88	87	92.91%	93	93	93
	9	63.21%	64	64	64	78.52%	79	80	78

Note: The **bold** numbers are the optimal results achieved.

TABLE 9.2

Selected Hyperparameters for the MLP and the KNN Classifier

MLP Classifier

Activation	ReLU
Solver	SGD
Alpha	1.00E-04
Learning rate	Adaptive
Learning rate initialization	1.00E-03
Maximum no. of iteration	300
Tolerance	1.00E-04
Momentum	0.8
Beta_1	0.9
Beta_2	0.999
Epsilon	1.00E-08
Hidden units	5
Hidden layers in each node	2

KNN Classifier

Weights	Uniform
Algorithm	Ball Tree
Leaf size	30
Distance metric	Minkowski
Number of nearest neighbors (K)	5
Power parameter (p)	2

optimization techniques to get the best feature set, and it is quite evident that if the feature space is vast, the performance of the entire model is improved by many folds.

9.4.4 OPTIMIZATION

By concatenation, we achieved an entire feature space of size 480×959, as we had 480 sample data and 959 features were extracted from each sample data. According to Özseven [30], feature selection can boost the classification result by quite a large margin, hence we have relied upon a novel feature selection in the acoustic emotion recognition task, named Manta Ray optimization. We have also experimented with three other popular optimization algorithms for feature selection. The brief descriptions of the methods are given next.

9.4.4.1 Manta Ray Optimization

Previously we have discussed that the MRFOA is constructed on three different optimization techniques such as chain foraging, cyclone foraging, and somersault foraging. Here in our current work, we have used two foragings cascaded for optimal feature selection. Initially, either chain or cyclone foraging is selected randomly, and using that particular technique, better a feature set among the entire feature set is achieved. Then, that particular feature set is again delivered as the input to the next and last foraging technique left, somersault foraging, for further optimal feature selection. The quantitative measurements are done by using the following equations:

$$Accuracy_i = \frac{\sum_i M_{ii}}{\sum_i \sum_j M_{ij}} \qquad (9.16)$$

$$Precision_i = \frac{M_{ii}}{\sum_j M_{ji}} \qquad (9.17)$$

$$Recall_i = \frac{M_{ii}}{\sum_j M_{ij}} \qquad (9.18)$$

$$F1\ score_i = \frac{1}{\dfrac{1}{Precision_i} + \dfrac{1}{Recall_i}} \qquad (9.19)$$

where M_{ij}= the weighted element of the confusion matrix at the ith row and j^{th} column. M_{ii} = the weighted diagonal element of the confusion matrix. We have also performed some comparative analysis of the MRFOA with some of the traditional feature selection algorithm to demonstrate the supremacy of the proposed algorithm over the other ones. Some of the algorithms, selected for comparison, are listed next.

9.4.4.2   Genetic Algorithm

Genetic algorithm is an evolutionary optimization that mimics the evolution process of the animal kingdom over generations to reach the global minima of a function that has the population as the solution. In general, the genetic algorithm is a very robust algorithm, which starts from any local minima of the solution function and usually reaches the global minima with the progression of generations.

We have initially selected a random subset of the entire feature space, known as a chromosome. The chromosome is represented as a binary vector having a dimension that is the same as that of the number of features in the entire feature space. In the chromosome, each feature vector is termed as a gene. In the binary vector or chromosome, the included or elected features were represented by 1, and the discarded features were represented by 0. Thus, some specific numbers of chromosomes were generated randomly and we formed a single population from such a stack of chromosomes. From the population, for each chromosome, a fitness value was determined, which is usually the prediction accuracy calculated using some classifier over the validation set generated from that particular feature subset or chromosome. Thereafter we selected out the chromosomes, having good fitness values from the entire population, and sent them to the mating pool for crossover. Crossover can be done in various ways, but we bisected each chromosome from the middle and exchanged the tail half portion of one chromosome with the other, which produced a single chromosome of the new generation. Following this aforementioned technique, the new generation population was generated from the old population. Another randomization is also introduced in the algorithm— mutation. Usually, mutation is a probabilistic hyperparameter determined by the user. Mutation exchanges a single gene from a chromosome with another gene from any randomly selected chromosome. Thus, the algorithm approached global minima and was able to find the fittest chromosome to produce the best classification results.

9.4.4.3 Particle Swarm Optimization

A particle swarm optimization (PSO) algorithm is a meta-heuristic algorithm based on the paradigm of swarm intelligence and is inspired by the social behavior of fish and birds. This algorithm presumes the local minima and global minima of a function are the various depths of a lake. The PSO algorithm is based on elementary techniques as communication and learning. Similar to the genetic algorithm, it also contains a population of randomly generated candidate solutions termed as a swarm. Each solution in the swarm is called a particle, which is also a randomly chosen subset of the entire feature space. $x_i(n)$ is the position of particle i with n the discrete time step. But another parameter velocity $v_i(n)$ of particle i has the same dimension as that of $x_i(n)$. The velocity vector denotes the movement of particle i in the sense of direction and distance concerning step size. Along with this, every particle has its memory of best position or the best minima value it has reached, which is termed as the personal best, denoted by $P_i(n)$. Along with all these, there is a common best experience or $\Delta g(n)$ that is particle invariant because it is a global best experience for all particles in the swarm. According to this simple mechanism, we updated the position and the velocity of each particle over each iteration. The vector between the personal best and positional vector and between global best and positional vector can be expressed by the following equations:

$$\overrightarrow{a_i(n)} = \overrightarrow{P_i(n)} - \overrightarrow{x_i(n)} \qquad (9.20)$$

$$\overrightarrow{b_i(n)} = \overrightarrow{\Delta g(n)} - \overrightarrow{x_i(n)} \qquad (9.21)$$

Using the above two vectors along with the velocity vector, the particle moves to a new position in the next iteration, i.e. the movement of the particle is at the direction of the resultant of these three vectors and that becomes the newly updated position of the particle in that iteration. It is given by the following equation:

$$\overrightarrow{P_i(n)}^N = \overrightarrow{a_i(n)} + \overrightarrow{b_i(n)} + \overrightarrow{v_i(n)} \qquad (9.22)$$

In the next iteration, the position was again updated. In the end, when the position of all the particles meets at a single point, it can be said that the solution has reached the global minima. That is how the particle swarm optimization algorithm was used for the optimal feature selection.

9.4.4.4 Grey Wolf Optimization

The grey wolf optimizer (GWO) is an optimization algorithm based on the behavior and social hierarchy of grey wolves. Wolves are categorized into four groups named alpha, beta, gamma, and delta. The group is dominated by alphas, betas are the decision-makers, and the lowest-ranked in the social hierarchy are deltas. The rest of them are gamma. The main components of GWO are encircling the prey, hunting, and attacking the prey. Initially, we began with some randomly generated agents and kept updating their position around the optimal solution (prey), known as encircling the prey. The position can be updated with the following equations:

$$\vec{d} = \left| \vec{a} \times \overrightarrow{p(t)} - \overrightarrow{g(t)} \right| \tag{9.23}$$

$$\overrightarrow{g(t+1)} = \overrightarrow{g(t)} - \vec{x} \times \vec{d} \tag{9.24}$$

where $\vec{a}$ and $\vec{x}$ are coefficient vectors, $\overrightarrow{p(t)}$ indicates the vector position of the prey, $\vec{g}$ indicates the vector position of the grey wolf. The best three agents who have better knowledge and updated position near the position of the prey are considered as the best solution and named as alpha, beta, and delta agents. The remaining agents are forced to update their positions around those three wolves to approach the best solution, known as hunting. The hunting process is represented by the following equations:

$$\overrightarrow{d_\alpha} = \left| \overrightarrow{a_1}.\overrightarrow{p_\alpha} - \vec{g} \right| \tag{9.25}$$

$$\overrightarrow{d_\beta} = \left| \overrightarrow{a_1}.\overrightarrow{p_\beta} - \vec{g} \right| \tag{9.26}$$

$$\overrightarrow{d_\delta} = \left| \overrightarrow{a_1}.\overrightarrow{p_\delta} - \vec{g} \right| \tag{9.27}$$

$$\overrightarrow{g_1} = \overrightarrow{g_\alpha} - \overrightarrow{x_1}.\overrightarrow{d_\alpha} \tag{9.28}$$

$$\overrightarrow{g_2} = \overrightarrow{g_\beta} - \overrightarrow{x_2}.\overrightarrow{d_\beta} \tag{9.29}$$

$$\overrightarrow{g_3} = \overrightarrow{g_\delta} - \overrightarrow{x_3}.\overrightarrow{d_\delta} \tag{9.30}$$

$$\overrightarrow{g(t+1)} = \frac{\overrightarrow{g_1} + \overrightarrow{g_2} + \overrightarrow{g_3}}{3} \tag{9.31}$$

9.5 RESULTS

Previously mentioned performance determining parameters of two different classifiers—KNN and MLP—on the previously discussed two datasets are described next.

TABLE 9.3

Quantitative Evaluation Results of KNN Classifier Using 5-fold Validation on SAVEE Dataset

Fold	Accuracy	Precision	Recall	F1 Score
Fold 1	96.56%	97%	97%	97%
Fold 2	95.26%	96%	96%	95%
Fold 3	97.81%	98%	98%	98%
Fold 4	96.56%	97%	97%	97%
Fold 5	96.56%	97%	97%	97%
Mean ± STD:	96.55% ± 0.80%	97.00% ± 0.63%	97.00% ± 0.63%	96.8% ± 0.98%

TABLE 9.4

Quantitative Evaluation Results of MLP Classifier Using 5-fold Validation on SAVEE Dataset

Fold	Accuracy	Precision	Recall	F1 Score
Fold 1	97.91%	98%	98%	98%
Fold 2	97.91%	98%	98%	98%
Fold 3	96.88%	97%	97%	97%
Fold 4	96.88%	97%	97%	97%
Fold 5	97.91%	98%	98%	98%
Mean ± STD:	**97.49% ± 0.50%**	**97.6% ± 0.48%**	**97.6% ± 0.48%**	**97.6% ± 0.48%**

9.5.1 RESULTS ON SAVEE DATASET

Our proposed method achieved an average classification accuracy of 96.55% with a standard deviation of ±0.80% and average accuracy of 97.49% with a standard deviation of ±0.50% using the MLP classifier. Tables 9.3 and 9.4 display the evaluation of our method using KNN and MLP classifiers, respectively. The bold results on Table 9.4 are the best results achieved by the MLP classifier. Table 9.7 is a comparative study of the recently evolved methods with our proposed method that prove our proposed method outperformed all the existing models and has achieved a state-of-the-art result on the SAVEE dataset.

9.5.2 RESULTS ON EmoDB DATASET

Our proposed method achieved an average accuracy of with standard deviation of on EmoDB dataset using the KNN classifier using five-fold validation and average accuracy of with standard deviation of using MLP classifier. Tables 9.5 and 9.6 are the quantitative measurements of our proposed approach on the EmoDB dataset using KNN and MLP classifiers, respectively. Table 9.8 shows a comparative study

TABLE 9.5

Quantitative Evaluation Results of KNN Classifier Using 5-fold Validation on EmoDB Dataset

Fold	Accuracy	Precision	Recall	F1 Score
Fold 1	96%	97%	96%	96%
Fold 2	95.43%	96%	96%	96%
Fold 3	95.96%	96%	96%	96%
Fold 4	97.21%	98%	98%	97%
Fold 5	95.43%	96%	96%	96%
Mean ± STD:	**96.06% ± 0.65%**	**96.60% ± 0.80%**	**96.40% ± 0.80%**	**96.2% ± 0.40%**

of our proposed approach with the recently evolved and existing methods on this dataset. The table shows that our method has achieved state-of-the-art results on the EmoDB dataset.

It is evident from the experimental results that our proposed method outperformed the existing methods, though we have used similar preprocessing and feature extraction methods. The method stands unique due to the application of meta-heuristic Manta Ray optimization that proved to be very efficient and robust in the feature selection task. The improved performance of the optimizer also depends upon the efficient selection of parameters which is described previously. This problem was solved by initializing the variables stochastically and then approaching the optimal value through small changes gradually.

The receiver operating characteristics (ROC) prove that the selected features are so well selected that the area under the curve for all the seven classes is close to 1, which also justifies the authenticity of the improved performance of the proposed method. An important observation in this aspect is that the MLP classifier performed slightly better than the KNN classifier in both cases with very few exceptions throughout the experiment. This can be justified by the nonparametric and random

TABLE 9.6

Quantitative evaluation results of MLP classifier using 5-fold validation on EmoDB dataset

Fold	Accuracy	Precision	Recall	F1 Score
Fold 1	97.25%	98%	97%	98%
Fold 2	98.31%	99%	99%	98%
Fold 3	98.31%	99%	99%	99%
Fold 4	98.31%	99%	99%	99%
Fold 5	96.22%	97%	97%	96%
Mean ± STD:	**97.68% ± 0.83%**	**98.40% ± 0.80 %**	**98.20% ± 0.97%**	**98.00% ± 1.09%**

TABLE 9.7

Comparison of Manta Ray Optimization with Other Feature Selection Algorithms in Terms of Accuracy, Precision, Recall, and F1 Score

Dataset	Optimization Algorithm	Number of Selected Features	Accuracy	Precision	Recall	F1 Score
SAVEE dataset	Genetic algorithm	325	72.23%	73%	73%	74%
	Grey wolf optimization algorithm	89	84.31%	85%	87%	82%
	Particle swarm optimization	35	81.32%	81%	82%	83%
	Manta ray optimization	43	97.49%	97%	98%	99%
EmoDB dataset	Genetic algorithm	230	85.91%	87%	86%	87%
	Grey wolf optimization algorithm	76	82.21%	83%	82%	83%
	Particle swarm optimization	26	87.29%	88%	87%	89%
	Manta ray optimization	61	97.68%	99%	98%	98%

approach as compared to the parametric and streamlined approach of MLP classifier. Though KNN has fewer parameters and easy to train, it has huge computational requirements and very slow in the case of large datasets.

9.5.3 COMPARISON WITH OTHER OPTIMIZATION ALGORITHMS

To prove the superiority of the Manta Ray optimization algorithm in this particular task, we have compared our result to the result achieved using three different optimization algorithms: (1) Genetic algorithm, (2) Grey Wolf Optimization, and (3) Particle Swarm Optimization. Although the results from these are significantly good, the results from our proposed method are far better than those obtained by these optimizations. Table 9.7 is a brief comparison of the results achieved by those two methods. The table also displays how the Manta Ray algorithm can perform better than the other algorithms, with very few percentages of selected features (second-best in the comparison table, after PSO for both the datasets). The probable reason behind this is supposed to be the two-fold optimization and their random initialization which can efficiently select the best features from a set of a huge number of features.

9.5.4 COMPARISON WITH EXISTING METHODS

Tables 9.8 and 9.9 display a comparative analysis of the obtained result with the existing results so far. The results are selected for comparison in such a way that the experimentation parameters (e.g. train-test splitting ratio on a particular dataset) are more or less similar to maintain uniformity. It is observed that in both cases, our method outperformed the existing approach by quite a large margin. Nguyen et al. [31] achieved the state-of-the-art results on the SAVEE dataset using PathNet,

TABLE 9.8
Comparative Analysis of the Obtained Result with the Existing Results

SL No.	Paper	Features	Method	Accuracy
1	Mao et al. [33]	CNN learns on its own	CNN	73.60%
2	Liu et al. [34]	MFCC	GA-BEL Model	76.40%
3	Nguyen et al. [31]	Neural Net learns by itself	PathNet	93.75%
4	Hajarolasvadi et al. [32]	Deep Features of proposed Neural Network	3D CNN-based approach using K-Means Clustering	81.05%
5	Pereira et al. [35]	Converted to Spectrogram via Short Time Fourier Transform (STFT)	Pre-Trained BEGAN with Libre Speech dataset was fine-tuned and used for evaluation.	40.05%
6	Our Proposed work	Concatenation of MFCC and LPC features	Optimal Feature Selection using Manta Ray Optimization Algorithm followed by classification with MLP Classifier and KNN classifier	Using KNN 96.55% Using MLP 97.49%

TABLE 9.9
Comparison of our Proposed Approach with the Recently Evolved Methodologies on the EmoDB Dataset Along with Their Brief Descriptions.

SL No.	Paper	Features	Method	Accuracy
1	Mao et al. [33]	CNN learns deep features automatically	CNN	85.20%
2	Deng et al. [22]	LLD features, like ZCR, RMS, energy, MFCC, frequency of pitch, HNR	Denoising Autoencoder	57.94%
3	Danisman et al. [37]	MFCC, F0, and total energy	SVM	63.52%
4	Albornoz et al. [38]	MFCC, Spectral, log spectrum, and prosodic features	SVM, HMM, Hierarchical classifier, GMM, and MLP	71.54%
5	Shen et al. [39]	LPCC, MFCC, LPCMCC, Energy, and pitch	SVM	82.57%
6	Wang et al. [36]	MFCC and Fourier parameters	SVM	88.88%
7	Wu et al. [16]	Speaking rate features, Prosodic features, features based on TEO and ZCR	SVM	91.03%
8	Our Proposed Work	Concatenation of MFCC, and LPC features	Optimal Feature Selection using Manta Ray Optimization Algorithm and classification with KNN and MLP Classifier	Using KNN 96.06% Using MLP 97.68%

which is a specialized neural network of its kind, designed especially for the classification of audio signals. The reported classification accuracy by Nguyen et al. [31] was 93.75% on the SAVEE dataset, achieved using the simplified approach of end-to-end feature extraction and classification. The reported result by Nguyen et al. [31] however, proved that CNN models can perform better than the machine learning classifiers, which is also justified by Hajarolasvadi et al. [32]. The researcher used 3D CNN models that utilize important spatio-temporal features that normal CNN often fails to use. The result reported in this chapter is 81.05% on the SAVEE dataset after using K-means clustering. Our proposed method achieved a classification accuracy of 97.49% using the MLP classifier and 96.55% using the KNN classifier on this particular dataset.

Table 9.9 compares the results obtained from previous existing methods with the one we achieved on the EmoDB dataset. Wu et al. [16] reported a state-of-the-art result on this dataset using the SVM classifier. The researcher extracted important features (e.g. features based on TEO and ZCR, prosodic features) using the traditional feature extraction methods and then used the SVM classifier to classify the acoustic dataset based on the extracted features, reporting a classification accuracy of 91.03%. Wang et al. [36], the researcher also obtained a classification accuracy of 88.88%, which is close to the previously mentioned result, by using MFCC and Fourier parameters. On the other hand, our method achieved a classification accuracy of 97.68% using MLP and 96.06% using KNN classifiers by using MFCC and LPC features from this dataset.

Figure 9.3 (4) represents the comparison of the evaluation parameters using MLP and KNN classifiers on EmoDB and SAVEE datasets. The bar graph shows that the MLP classifier performed better in both cases. However, none of these methods was based on optimization and feature selection algorithms, which can immensely contribute to the classification accuracy by selecting the most suitable features and discarding the redundant ones. Therefore, we can conclude that our model performed far better than the existing models due to increase in number of features extraction that often contains important information about the dataset, and then selecting the best ones using optimization.

9.6   CONCLUSION AND FUTURE WORK

From the experiments performed in these works, we reached some of the conclusions regarding the acoustic signal analysis and speech emotion recognition task. These conclusions are as follows:

- Feature extraction plays an important role in the classification task as they contain important information about the dataset. However, it is not possible to extract all the features, as it is a perennial task and requires lots of imaginations of the feature-engineer.
- This problem can be solved by using the feature concatenation method. This method is based on features extraction by different traditional methods and finally merging them; this can compensate for the missing information from the dataset.

- Selecting the optimal feature is an important step before classification and can contribute immensely to bridge the missing link between feature extraction and improved classification accuracy.
- Neural networks should be preferred for the classification of audio signals over other machine learning models. In our case, the MLP classifier performed better and faster than the KNN classifier in all the cases.

This chapter aims to contribute to the improvement of the SER task using a meta-heuristic MRFOA for discarding redundant features and selecting the most accurate ones for classification. The neural network-based MLP classifier achieved better in both the datasets as compared to the KNN classifier. This is also supported in Figure 9.3 (3) where the area under the ROC curve shows better results in MLP classifier in both the datasets. However, we aim for validating our model on other datasets, and also, we would like experiment with different optimization algorithms. Further, we would like to assess the model performance with variations in sentence length as it has been found difficult to classify long sentences than small syllables. Finally, we would like to perform experiments on other languages and try to evaluate the performance of native languages as there exists some relationship between accents and ease of classification.

REFERENCES

1. Bottesch T, Palm G. "Improving Classification Performance by Merging Distinct Feature Sets of Similar Quality Generated by Multiple Initializations of mRMR," *2015 IEEE Symposium Series on Computational Intelligence* (IEEE, 2015), pp. 328–334.
2. Maniar S, Shah JS. "Classification of Content-Based Medical Image Retrieval Using Texture and Shape Feature with Neural Network," *International Journal of Advances in Applied Sciences* 6, no. 4 (2017): 368–74.
3. Rong J, Li G, Chen YP. "Acoustic Feature Selection for Automatic Emotion Recognition From Speech," *Information Processing and Management* 45, no. 3 (May 2009), 315–28.
4. Rao KS, Koolagudi SG, Vempada RR. "Emotion recognition from speech using global and local prosodic features." *International Journal of Speech Technology* 16, no. 2 (June 2013), 143–60.
5. Stuhlsatz A, Meyer C, Eyben F, Zielke T, Meier G, Schuller B. "Deep Neural Networks for Acoustic Emotion Recognition: Raising the Benchmarks," *2011 IEEE International Conference on Acoustics, Speech and Signal Processing* (IEEE, May 2011), 5688–5691.
6. Kim J. "Bimodal Emotion Recognition Using Speech and Physiological Changes," *Robust Speech Recognition And Understanding* 265 (June 2007), 280.
7. Wu S, Falk TH, Chan WY. "Automatic Speech Emotion Recognition Using Modulation Spectral Features," *Speech Communication* 53, no. 5 (May 2011), 768–85.
8. Liu ZT, Wu M, Cao WH, Mao JW, Xu JP, Tan GZ. "Speech Emotion Recognition Based on Feature Selection and Extreme Learning Machine Decision Tree," *Neurocomputing* 273 (January 2018), 271–80.
9. Lalitha S, Patnaik S, Arvind TH, Madhusudhan V, Tripathi S. "Emotion Recognition Through Speech Signal for Human-Computer Interaction," *2014 5th International Symposium on Electronic System Design (IEEE, December 2014)*, 217–218.
10. Qin S, Song J, Zhang P, Tan Y. "Feature Selection for Text Classification Based on Part of Speech Filter and Synonym Merge," *2015 12th International Conference on Fuzzy Systems and Knowledge Discovery* (IEEE, August 2015), 681–685).

11. Zhang R, Yang Y. "Merging Recovery Feature Network to Faster RCNN for Low-Resolution Images Detection," *2017 IEEE Global Conference on Signal and Information Processing* (IEEE, November 2017), pp. 1230–1234.

12. Schuller B, Reiter S, Muller R, Al-Hames M, Lang M, Rigoll G. "Speaker Independent Speech Emotion Recognition by Ensemble Classification," *2005 IEEE International Conference on Multimedia and Expo* (IEEE, July 2005), 864–867.

13. Zhao W, Zhang Z, Wang L. "Manta Ray Foraging Optimization: An Effective Bio-Inspired Optimizer for Engineering Applications," *Engineering Applications of Artificial Intelligence* 87 (January 2020): 103300.

14. Nakatsu R, Nicholson J, Tosa N. "Emotion Recognition and Its Application to Computer Agents with Spontaneous Interactive Capabilities," *Multimedia '99: Proceedings of the 7th ACM international conference on Multimedia (Part 1),* October 1999, 343–351.

15. Schuller B, Rigoll G, Lang M. "Hidden MARKOV Model-Based Speech Emotion Recognition," *2003 IEEE International Conference on Acoustics, Speech, and Signal Processing* (IEEE, April 2003), 2: II-1.

16. Wu CH, Liang WB. "Emotion Recognition of Affective Speech Based on Multiple Classifiers Using Acoustic-Prosodic Information and Semantic Labels," *IEEE Transactions on Affective Computing* 2, no. 1 (December 2010), 10–21.

17. El Ayadi M, Kamel MS, Karray F. "Survey on Speech Emotion Recognition: Features, Classification Schemes, and Databases," *Pattern Recognition* 44, no. 3 (March 2011), 572–87.

18. Ververidis D, Kotropoulos C. "Fast and Accurate Sequential Floating Forward Feature Selection with the Bayes Classifier Applied To Speech Emotion Recognition," *Signal Processing* 88, no. 12 (December 2008), 2956–70.

19. Zhao J, Mao X, Chen L. "Speech Emotion Recognition Using Deep 1D\& 2D CNN LSTM Networks," *Biomedical Signal Processing and Control* 47 (January 2019), 312–23.

20. Lee CC, Mower E, Busso C, Lee S, Narayanan S. "Emotion Recognition Using a Hierarchical Binary Decision Tree Approach," *Speech Communication* 53, no. 9–10 (November 2011), 1162–71.

21. Nwe TL, Foo SW, De Silva LC. "Speech Emotion Recognition Using Hidden Markov Models," *Speech Communication* 41, no. 4 (November 2003), 603–23.

22. Deng J, Zhang Z, Marchi E, Schuller B. "Sparse Autoencoder-Based Feature Transfer Learning for Speech Emotion Recognition," *2013 Humaine Association Conference on Affective Computing and Intelligent Interaction* (IEEE, September 2013) 511–516.

23. Fayek HM, Lech M, Cavedon L. "Evaluating Deep Learning Architectures for Speech Emotion Recognition," *Neural Networks* 92 (August 2017), 60–8.

24. Bengio Y, Courville A, Vincent P. "Representation Learning: A Review and New Perspectives," *IEEE Transactions on Pattern Analysis and Machine Intelligence* 35, no. 8 (March 2013),1798–828.

25. Rozgić V, Ananthakrishnan S, Saleem S, Kumar R, Prasad R. "Ensemble of SVM Trees for Multimodal Emotion Recognition," *Proceedings of the 2012 Asia Pacific Signal and Information Processing Association Annual Summit and Conference* (IEEE, December 2012), 1–4.

26. Sun Y, Wen G, Wang J. "Weighted Spectral Features Based on Local Hu Moments for Speech Emotion Recognition," *Biomedical Signal Processing and Control* 18 (April 2015), 80–90.

27. Lanjewar RB, Mathurkar S, Patel N. "Implementation and Comparison of Speech Emotion Recognition System Using Gaussian Mixture Model (GMM) and K-Nearest Neighbor (K-NN) Techniques," *Procedia Computer Science* 49 (January 2015), 50–7.

28. Kaya H, Karpov AA. "Efficient and Effective Strategies for Cross-Corpus Acoustic Emotion Recognition," *Neurocomputing* 275 (January 2018), 1028–34.

29. Mirjalili S, Saremi S, Mirjalili SM, Coelho LD. "Multi-Objective Grey Wolf Optimizer: A Novel Algorithm for Multi-Criterion Optimization," *Expert Systems with Applications* 47 (April 2016), 106–19.
30. Özseven T. "A Novel Feature Selection Method for Speech Emotion Recognition," *Applied Acoustics* 146 (March 2019), 320–6.
31. Nguyen D, Nguyen K, Sridharan S, Abbasnejad I, Dean D, Fookes C. "Meta Transfer Learning for Facial Emotion Recognition," *2018 24th International Conference on Pattern Recognition* (IEEE, August 2018), 3543–3548.
32. Hajarolasvadi N, Demirel H "3D CNN-Based Speech Emotion Recognition Using K-Means Clustering and Spectrograms," *Entropy* 21, no. 5 (May 2019), 479.
33. Mao Q, Dong M, Huang Z, Zhan Y. "Learning Salient Features for Speech Emotion Recognition Using Convolutional Neural Networks." *IEEE Transactions on Multimedia* 16, no. 8 (September 2014), 2203–13.
34. Liu ZT, Xie Q, Wu M, Cao WH, Mei Y, Mao JW. "Speech Emotion Recognition Based on an Improved Brain Emotion Learning Model," *Neurocomputing* 309 (October 2018), 145–56.
35. Pereira I, Santos D, Maciel A, Barros P. "Semi-Supervised Model for Emotion Recognition in Speech," *International Conference on Artificial Neural Networks* (Springer, Cham., October 2018), 791–800).
36. Wang K, An N, Li BN, Zhang Y, Li L. "Speech Emotion Recognition Using Fourier Parameters," *IEEE Transactions and Affective Computing* 6, no. 1 (January 2015), 69–75.
37. Danisman T, Alpkocak A. "Emotion Classification of Audio Signals Using Ensemble of Support Vector Machines," *International Tutorial and Research Workshop on Perception and Interactive Technologies for Speech-Based Systems* (Springer: Berlin, Heidelberg, June 2008), 205–216.
38. Albornoz EM, Milone DH, Rufiner HL. "Spoken Emotion Recognition Using Hierarchical Classifiers," *Computer Speech and Language* 25, no. 3 (July 2011), 556–70.
39. Shen P, Changjun Z, Chen X. "Automatic speech emotion recognition using support vector machine," *Proceedings of 2011 International Conference on Electronic & Mechanical Engineering and Information Technology* (IEEE, August 2011), 2: 621–625.

10 Internet of Things in Healthcare Informatics

A. Vijayalakshmi and A. Hridya

CONTENTS

10.1 INTRODUCTION

When the term "Internet of Things" (IoT) was coined by Kevin Ashton in 1999, he envisioned computers to empower with their own capability to gather data from surroundings. IoT is reconstructing the current world into a dynamic world of interconnected devices in a very exponential way. IoT is a spectacular phenomenon that enables to connect any one at any point. There is a rapid increase in number of

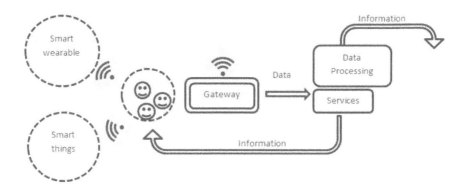

FIGURE 10.1   Data processing in IoT.

such compact smart devices that can exchange data [1]. With this adaptation of new technologies, IoT is well accepted by various industries. The major advancements are in this field is in the aspect of network, communication and security. Figure 10.1 depicts the various components in an IoT environment with the flow of data from one unit to other.

In the foreseeable future, IoT will increase by integrating objects leading to a distributed network environment. This, in turn, will open wide opportunities for various novel applications with an objective to increase quality of life. It is expected that the growth of IoT devices may reach 50 billion by the end of 2020 [2]. The miniaturization of devices has been a great motivation to IoT, which in turn has increased the network bandwidth to a larger scale. IoT ecosystems contain various components to capture and transmit data. These primary components play a major role in the overall behavior of IoT in enabling a plethora of services:

1. **Sensors**: In the architecture of IoT, sensors are the major components in the bottom layer. The basic function of these end devices is to capture signals from the environment and transmit them for further processing. The data collected through these sensors are used in various fields of IoT environments. The lowest layer connects to the network layer.
2. **Connectivity**: Connecting to the internet is a crucial part in IoT. Network connectivity can be chosen according to the use case, but options vary from WiFi, Bluetooth, and RFID, etc. The connectivity option varies between the consumption of power, range, and bandwidth of the network. It takes more power to transfer data over a long distance, so to decrease the power consumed by the devices, the range can, in turn, be shortened.
3. **Data processing**: There is an extraordinary amount of data collected from devices in an IoT environment. Cloud solutions are a great benefit for the amount of storage space required to process this huge capacity of data. Data processing is the most challenging and crucial component of IoT solutions. In the data processing stage, different practices like classification and prediction are used to get information from the data received.

4. **User Interface**: Displaying the result of analyzed IoT data to a user should be understandable so to visualize the change in continuous stream of data. The massive amount of data collected and further processed in IoT is delivered to the end user through various interfaces like mobile apps, smart devices, or voice-controlled devices.

10.2   ARCHITECTURE OF INTERNET OF THINGS

There is no general architecture for Internet of Things that is agreed upon by all the researchers [2]. Different architectures are proposed by researchers based on the objectives of the applications. This chapter discusses the four-layer architecture IoT of as depicted in Figure 10.2. Four-layer architecture was proposed because three-layer architecture was not fulfilling the requirements of IoT.

10.3   THE ROLE OF IoT IN HEALTHCARE

Utilizing IoT in health care is very perplexing because of cost effectiveness and use of many accurate medical sensors in this area. Also, since the sensors are of heterogeneous nature and much data is generated out of them, involving IoT in health care seems like a great challenge. But this is possible by exploring algorithms and other concepts of IoT to make health care a smart system [3]. For this purpose, wireless technologies have been integrated so that the devices can be monitored as well

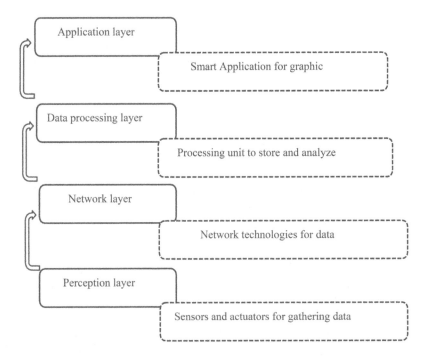

FIGURE 10.2   IoT architecture for health care.

as act as network managers. This will help in connecting all healthcare resources and integrate them with hospitals, rehabilitation centers, medical professionals, and ambulances. The server needs to be centralized and it should be capable of performing data analysis, managing critical events, and also developing new strategies based on current situations. To make IoT health care smarter, some devices not capable of becoming integrated into the network should be made a part of it so that the effectiveness of the system can be increased [4]. The uses of mobile devices with the integration of IoT is on the rise as this has enabled healthcare facilities to collect patient data in real time, which has not only changed the system in a better way but has brought a revolution to medical systems. This is also called m-health. The main advantage of the integration of m-health and IoT has made it easy to obtain a way to communicate with the sensors, which are utilized to collect real-time information. The devices used for this purpose are termed smart devices, which can monitor many parameters, such as blood sugar levels, blood pressure, and heart rate. For this purpose, many algorithms are used so that data can be analyzed in real time to obtain certain patterns in its growth, which allows the ability of generating an alert if any irregular activities are noticed, or in case of emergency [5].

The role of IoT in health care is vast, as IoT is applied in almost all aspects of the healthcare industry ranging from clinical care to inventory. Treating a patient usually requires conducting tests like blood, urine, etc. These tests do provide accurate results, but the major disadvantage is that the tests may take time to provide results, which can cause delays in treatment. By introducing the concept of IoT, a quick and pervasive way of diagnosing the patients during an emergency situation can be carried out as real-time information such as glucose level, blood pressure level, cardiac routine, etc. can be obtained [6]. The second aspect is IoT's utilization in remote monitoring, which can easily help medical professionals obtain patient data and information anywhere and anytime [6, 7]. This will help doctors to assist patients in emergency cases. The third aspect is IoT's use in the robotics and medical fields. Robots are used to perform rigorous and strenuous activities in the absence of medical professionals. Robots in the medical field can be utilized to perform complex operations, monitor patients, and care for patients in rehabilitation centers. Apart from these, IoT can be used in home settings to monitor a patient's quality of air, temperature, and cardiac activity [8] and track medicine intake [7].

The increase in population has not only brought scarcity in resources but also has led to many people not being able to utilize medical resources. Introducing IoT into the medical field not only provides resource utilization, but will also allow physicians to effectively monitor patients anywhere and at any time.

10.4 SERVICES OF IOT IN HEALTHCARE SYSTEMS

The term IoT is a phenomenon where in the physical components are interconnected for a fast response by exchanging data sensed from the environment. This technology has enabled the healthcare industry to improve communication with various stakeholders for quick medical assistance. It has also added value in monitoring remote patients and providing treatment [9]. Various services in IoT that have contributed to the healthcare industry are listed next.

10.4.1 SERVICE TO PATIENTS

Integration of IoT in various devices has changed the life of the modern population. This has been a great advantage when the integration of miniature devices is possible in wearable bands to monitor various activities. Many of these devices also help in setting reminders, which can be of a great help to elderly patients or caretakers.

10.4.2 SERVICE TO PHYSICIANS

Integration of user interfaces has been an advantage for physicians to monitor patient. The data collected from patients based on their daily activities is a great advantage when relayed in a timely manner to the physician, allowing him/her to provide effective support.

10.4.3 SERVICE TO HOSPITALS

At hospitals, IoT has contributed a plethora of services ranging from monitoring patients and their daily activities to keeping track of the inventory of medicines. IoT can also keep track of equipment in the hospitals for easy access.

10.4.4 SERVICE TO ELDERLY

Ambient Assisted Living (AAL) is a well-received IoT platform in an era where the populations of older adults living alone are increasing. This technology, driven by AI, helps in monitoring the health conditions of the elderly and gives appropriate indications to the healthcare stake holders. IoT-enabled devices in AAL help the elderly living alone to be safe and receive assistance in health-related matters. AAL services provide assistant to individuals similar to a caretaker. The advantage in the case of IoT-enabled service is that the sensors collect real-time information and analyzes the data to provide effective services [10].

10.4.5 SERVICES THROUGH M-HEALTH DEVICES

m-Health technology is an advancement in medical field to provide effective communication in real time to avail medical services. m-Health involves mobile devices that play an important role in collecting health-related data from patients and helps in storing this data in servers to be available for various healthcare stake holders. The collected data can be used by doctors, caretakers, or relatives for monitoring and providing assistance when necessary. Using such devices can be of great benefit in identifying the discrepancies in real time [11].

10.5 APPLICATION OF IOT IN HEALTHCARE

It can be a serious condition for elderly who live alone when they need medical care in short intervals. The time and money spent in a hospital is not a possibility for a large part of the population. Traditional healthcare systems have various medical devices that are specific to individual problems and are available only in hospitals or clinics. For each of the minor health-related issues, patients from remote places have to travel and visit hospitals or medical clinics for their periodic checkups. The integration of IoT in healthcare has given a helping hand in such situations where various smart devices are developed that can assist patients [12]. The various applications where IoT is integrated in healthcare are explained next [9].

10.5.1 Sensing Glucose Level

Diabetes is a common health issue caused by increased blood glucose (also known as blood sugar). There are various medical IoT-enabled real-time applications that are designed to sense glucose levels without disturbing the patient. Monitoring blood glucose levels helps in identifying the patterns of blood glucose, which can be used in planning for appropriate food consumption in diabetes patients. These devices are connected to various end-user applications to give real-time suggestions to healthcare providers [13].

10.5.2 Monitoring ECG

Electrocardiogram (ECG) monitoring is a very important parameter in diagnosing heart-related diseases. Applying IoT in ECG monitoring has been a great advantage to users in receiving medical attention at appropriate instances through real-time analysis of data. An IoT-based ECG monitoring system consists of a portable wireless transmitter and receiver that can detect abnormal data in real time. The application for the end-user contains mobile applications for receiving indications from the server based on the analysis of the data.

10.5.3 Monitoring Blood Pressure

The availability of wearable devices that help monitor health is easily available and well accepted. Wearable devices to monitor blood pressure are a combination of blood pressure meter and smartphone based on the concept of IoT. Blood pressure, in this case, is continuously monitored and any changes that fall beyond the normal values are indicated to healthcare professionals in real time, which can save lives.

10.5.4 Body Temperature Monitoring

One of the most evident parameters that can indicate a change in the health of a person is body temperature. Monitoring body temperature will be advantageous for healthcare professionals to detect significant changes in the body. medical Internet of Things(m-IoT)–based systems that can detect a change in temperature on a real-time basis are an added advantage to the healthcare industry. To design this, a temperature sensor is embedded in the device design to collect data and transfer it to the edge device or a server. The data analyzed will be sent to the end devices like mobile applications for easy user interface.

10.5.5 Monitoring Oxygen

To monitor the concentration of oxygen in the blood, pulse oximetry is most suitable. An IoT-based application to monitor oxygen level is advantageous for healthcare applications. There are various IoT-based wearable pulse oximeters proposed for healthcare applications. Such devices with various types of connectivity can send data to a server. IoT-based oxygen monitoring systems can monitor patients remotely to identify the change in temperature through wearable devices. Such devices can be used to monitor the temperature of a person/patient continuously.

10.5.6 Cancer Care

Cancer originates from tumors and there are two types of tumors: benign and malignant. Malignant tumors are responsible for the rapid multiplication of cells, which leads to various patient complications including death. Therefore, by applying IoT technologies and its services, cancer can be detected, diagnosed, and also be treated. These smart services can be implemented by using sensor technologies that can measure different parameters for cancer care services and provide the right diagnosis and detection in case of emergencies. Also, with the use of cancer care applications, counseling and support can be provided to the patients along with analyzing and viewing the reports in due time [14].

10.5.7 Anomaly Detection

Healthcare data is of utmost importance for human survival. Detection of various diseases at a very early stage is important as this can prevent many undesirable outcomes. Anomaly detection helps to detect many such diseases with low false-positive rates. This can be used to view and maintain the electronic health records of patients, medical analysis of images, biomedical signal analysis, and also local analysis by using big data mining and pattern recognition algorithms, such as association rule mining and pattern mining [15].

10.5.8 Cardiovascular Diseases

IoT is often known as an efficient decision-making system and also a global healthcare platform capable of fulfilling goals. This can be applied to find out the risk groups capable of developing cardiovascular diseases. Based on the data obtained, which can be manually entered by doctors and patients, risk groups are classified as risky and nonrisk categories. Also by this data, prescriptions can be updated by doctors using the applications [16].

10.5.9 Human Activity Recognition

The nature of health has gradually changed because of the environment and the activities of people in their daily lives. By the integration of IoT and smartphones, the health and activity of a patient can be monitored and, in case of emergency or variations in the activity, the patient can be alerted. The sensors are integrated to various wearable devices to capture data from ankle, chest, wrist etc. Using these wearable devices allows information about human activity, such as cycling, walking, and sleeping, to be monitored and analyzed using artificial neural network algorithms [17].

10.5.10 REMOTE MONITORING OF PATIENTS

Remote patient monitoring, also known as telehealth, allows patients to conduct their routine tests and send real-time data to the healthcare providers through a software application that will be installed on a smartphone. A few examples are glucose meters for diabetic patients and blood pressure and heart monitoring for cardiac care. This can drastically reduce the amount of time a patient spends visiting the hospital [18].

10.6 PROCESSING THE IoT DATA

The volume of data generated from IoT is enormous and there should be a mechanism derived to manage and analyze it [19]. Figure 10.3 demonstrates the schematic representation of processing IoT data. The data generated from sensors is transferred via various networking devices to cloud servers for processing through big data technologies [20].

IoT is defined as the environment where smart devices are deployed and collect data. These devices are connected to each other and communicate to transfer

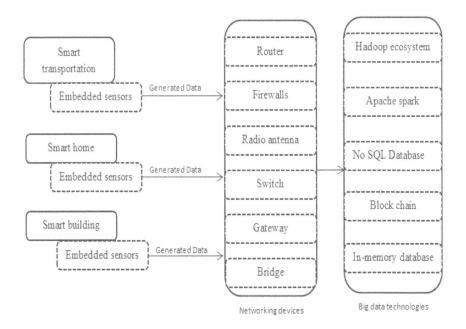

FIGURE 10.3   IoT components and data science.

information. Data science plays an important role in extracting this information for prediction and decision making.

10.7 ANALYZING THE IoT DATA

Data collected from IoT devices are enormous and data analysis tools can be applied to these various types of data to derive knowledge [19]. Valuable information can be derived using several types of IoT analytics as described below:

1. **Perspective analytics**: Used to analyze the steps to be taken for a certain situation. It enables the best action to be taken at a current situation. Perspective analysis is often described as a combination of descriptive and predictive analysis.
2. **Descriptive analytics**: This simple form of analytics helps describe and summarize the IoT data. The data collected are interpreted and then presented to be understood based on the context.
3. **Predictive analytics**: This type of analytic uses past data for predicting future results and it will have various methods to calculate the quality of these predictions.
4. **Diagnostic analytics**: Diagnostic analytics finds the root cause for IoT data and is based on statistical models that examine the relationship between data. It helps in understanding why a particular thing has happened with IoT data and if a particular alert is valid.
5. **Time series analytics**: This is based on time-based data. Health monitoring and weather monitoring are two domains where time series data comes into picture. These data are analyzed to find any anomalies.

10.8 SECURITY CHALLENGES WITH HEALTHCARE DATA

Internet of Things (IoT) is an interconnection of physical devices for exchanging data. IoT has been positively received by hospitals for many years [18]. IoT devices are common in the hospitals as well as in palliative care centers. In the healthcare domain, sharing personal data is a major concern and it can affect individuals directly or indirectly [21]. There are various advantages of connected devices in the IoT environment; however the risk on privacy of data has to be a primary concern. Some of the risks associated with this environment are: unauthorized access that leads to misuse of personal data, facilitates attacks on other systems in the network, risk on personal safety, and privacy risks. In 2015, it was found that thousands of medical devices like anesthesia equipment, various cardiology devices, etc. were exposed to various vulnerabilities as they were connected to the internet [22]. Such devices are easily attacked to collect data by brute force attack and hard-coded logins. Hence, security and privacy are major concerns when data are transmitted across the network [23]. The US Department of Health and Human Service has introduced HIPAA (the Health Insurance Portability and Accountability Act of 1996) [25] for protecting the privacy and security of health information collected of patients. In IoT architecture, security is to be maintained at every layer. Authentication of devices and providing access to controls are most important to security considerations.

10.8.1 MEASURES TO OVERCOME SECURITY ISSUES IN IoT

Various IoT applications are accepted by the healthcare industry and there is an exponential growth in this field. Sharing of real-time health information has great efficiency in fast and accurate response and saves time and costs. Security of these data is crucial and many manufactures of IoT devices use encryption and secure boot features, which makes sure that when the device is started, none of its configurations can be changed. These security measures have to be incorporated and assessed before the device is released into the market [24]:

1. Only authorized personal should be given access to the device.
2. Incorporate extra authentication measures like biometrics that can reduce password attack.
3. Deployment of automated threat intelligence measures that can continuously monitor the network and protect the system from various threats like password-based attacks.
4. Monitoring of device-to-device communication
5. Limit the access to the device.
6. Verify the firmware that is sent to the device.
7. Different layers in the architecture should be secured against various threats.
8. Security of data on the device should be monitored continuously.
9. Implement security analytics that involve collecting and correlating data from multiple sources and analyze them to identify various threats.
10. Employees should be continuously educated on best practices for creating passwords.

10.9 CONCLUSION

IoT-enabled devices are growing tremendously, providing various new capabilities to the world. Health care is one such domain where integrating networking capability in various equipment to provide continuous monitoring and services is such a great advantage. This chapter discusses various services and applications of IoT in health care. This envisions the services of IoT that facilitate interaction between various stakeholders. Finally, the challenges of IoT and the measures to avoid various associated risks are highlighted, which can open the road to research in the domain.

REFERENCES

1. Miorandi, Daniele, Sabrina Sicari, Francesco De Pellegrini, and Imrich Chlamtac. 2012. "Internet of Things: Vision, Applications and Research Challenges." *Ad Hoc Networks* 10 (7): 1497–1516. https://doi.org/10.1016/j.adhoc.2012.02.016.

2. Burhan, Muhammad, Rana Rehman, Bilal Khan, and Byung-Seo Kim. 2018. "IoT Elements, Layered Architectures and Security Issues: A Comprehensive Survey." *Sensors* 18 (9): 2796. https://doi.org/10.3390/s18092796.

3. Qi, Jun, Po Yang, Geyong Min, Oliver Amft, Feng Dong, and Lida Xu. 2017. "Advanced Internet of Things for Personalised Healthcare Systems: A Survey." *Pervasive and Mobile Computing* 41 (October): 132–49. https://doi.org/10.1016/j.pmcj.2017.06.018.

4. YIN, Yuehong, Yan Zeng, Xing Chen, and Yuanjie Fan. 2016. "The Internet of Things in Healthcare: An Overview." *Journal of Industrial Information Integration* 1 (June): 3–13. https://doi.org/10.1016/j.jii.2016.03.004.

5. Almotiri, Sultan H., Murtaza A. Khan, and Mohammed A. Alghamdi. 2016. "Mobile Health (m-Health) System in the Context of IoT." *Proceedings – 2016 4th International Conference on Future Internet of Things and Cloud Workshops, W-FiCloud 2016*, 39–42. https://doi.org/10.1109/W-FiCloud.2016.24.

6. Pramanik, Pijush Kanti Dutta, Bijoy Kumar Upadhyaya, Saurabh Pal, and Tanmoy Pal. 2019. "Internet of Things, Smart Sensors, and Pervasive Systems: Enabling Connected and Pervasive Healthcare." *Healthcare Data Analytics and Management*. London: Elsevier Inc. https://doi.org/10.1016/b978-0-12-815368-0.00001-4.

7. Zeadally, Sherali, and Oladayo Bello. 2019. "Harnessing the Power of Internet of Things Based Connectivity to Improve Healthcare." *Internet of Things*, no. xxxx: 100074. https://doi.org/10.1016/j.iot.2019.100074.

8. Nazir, Shah, Yasir Ali, Naeem Ullah, and Iván García-Magariño. 2019. "Internet of Things for Healthcare Using Effects of Mobile Computing: A Systematic Literature Review." *Wireless Communications and Mobile Computing* 2019. https://doi.org/10.1155/2019/5931315.

9. Riazul Islam, S. M., Daehan Kwak, Md Humaun Kabir, Mahmud Hossain, and Kyung-Sup Kwak. 2015. "The Internet of Things for Health Care: A Comprehensive Survey." *IEEE Access* 3: 678–708. https://doi.org/10.1109/ACCESS.2015.2437951.

10. Dohr, A., R. Modre-Opsrian, M. Drobics, D. Hayn, and G. Schreier. 2010. "The Internet of Things for Ambient Assisted Living." In *2010 Seventh International Conference on Information Technology: New Generations*, 804–9. Las Vegas, NV, USA: IEEE. https://doi.org/10.1109/ITNG.2010.104.

11. Dhanvijay, Mrinai M., and Shailaja C. Patil. 2019. "Internet of Things: A Survey of Enabling Technologies in Healthcare and Its Applications." *Computer Networks* 153 (April): 113–31. https://doi.org/10.1016/j.comnet.2019.03.006.

12. Abinaya, Vinoth Kumar, and Swathika. 2015. "Ontology Based Public Healthcare System in Internet of Things (IoT)." *Procedia Computer Science* 50: 99–102. https://doi.org/10.1016/j.procs.2015.04.067.

13. Chatterjee, Samir, Jongbok Byun, Kaushik Dutta, Rasmus Ulslev Pedersen, Akshay Pottathil, and Harry (Qi) Xie. 2018. "Designing an Internet-of-Things (IoT) and Sensor-Based in-Home Monitoring System for Assisting Diabetes Patients: Iterative Learning from Two Case Studies." *European Journal of Information Systems* 27 (6): 670–85. https://doi.org/10.1080/0960085X.2018.1485619.

14. Onasanya, Adeniyi, and Maher Elshakankiri. 2019. "Smart Integrated IoT Healthcare System for Cancer Care." *Wireless Networks*, January. https://doi.org/10.1007/s11276-018-01932-1.

15. Ukil, Arijit, Soma Bandyoapdhyay, Chetanya Puri, and Arpan Pal. 2016. "IoT Healthcare Analytics: The Importance of Anomaly Detection." *Proceedings - International Conference on Advanced Information Networking and Applications, AINA* 2016-May: 994–97. https://doi.org/10.1109/AINA.2016.158.

16. Chatterjee, Parag, Leandro J. Cymberknop, and Ricardo L. Armentano. 2018. "IoT-Based Decision Support System for Intelligent Healthcare – Applied to Cardiovascular Diseases." *Proceedings – 7th International Conference on Communication Systems*

and Network Technologies, CSNT 2017, 362–66. https://doi.org/10.1109/CSNT.2017.
8418567.

17. Subasi, Abdulhamit, Mariam Radhwan, Rabea Kurdi, and Kholoud Khateeb. 2018.
    "IoT Based Mobile Healthcare System for Human Activity Recognition." *2018 15th
    Learning and Technology Conference* (2018), 29–34. https://doi.org/10.1109/LT.2018.
    8368507.

18. Chacko, Anil, and Thaier Hayajneh. 2018. "Security and Privacy Issues with IoT in
    Healthcare." *EAI Endorsed Transactions on Pervasive Health and Technology* 0 (0):
    155079. https://doi.org/10.4108/eai.13-7-2018.155079.

19. Siow, Eugene, Thanassis Tiropanis, and Wendy Hall. 2018. "Analytics for the Internet
    of Things: A Survey." *ACM Computing Surveys,* 51 (4), 1–36. Availabel at: http://arxiv.
    org/abs/1807.00971.

20. Brohi, Sarfraz Nawaz, Mohsen Marjani, Ibrahim Abaker Targio Hashem, Thulasyammal
    Ramiah Pillai, Sukhminder Kaur, and Sagaya Sabestinal Amalathas. 2019. "A Data
    Science Methodology for Internet-of-Things." In *Emerging Technologies in Computing,
    Second International Conference for Emerging Technologies in Computing,* edited by
    Mahdi H. Miraz, Peter S. Excell, Andrew Ware, Safeeullah Soomro, and Maaruf Ali,
    285:178–86. Lecture Notes of the Institute for Computer Sciences, Social Informatics
    and Telecommunications Engineering series. Cham: Springer International Publishing.
    https://doi.org/10.1007/978-3-030-23943-5_13.

21. Andrea, Ioannis, Chrysostomos Chrysostomou, and George Hadjichristofi. 2015.
    "Internet of Things: Security Vulnerabilities and Challenges." In *2015 IEEE Symposium
    on Computers and Communication (ISCC),* 180–87. Larnaca: IEEE. https://doi.
    org/10.1109/ISCC.2015.7405513.

22. Cimpanu, Catalin. 2015. "Thousands of IoT Medical Devices Found Vulnerable to
    Online Attacks," September 29, 2015. https://news.softpedia.com/news/thousands-of-
    iot-medical- devices-found-vulnerable-to-online-attacks-493144.shtml.

23. Al-Issa, Yazan, Mohammad Ashraf Ottom, and Ahmed Tamrawi. 2019. "EHealth
    Cloud Security Challenges: A Survey." *Journal of Healthcare Engineering* 2019
    (September): 1–15. https://doi.org/10.1155/2019/7516035.

24. Gloss, Kristen. 2020. "Healthcare IoT Security Risks and What to Do about Them."
    August 14, 2020. https://internetofthingsagenda.techtarget.com/feature/Healthcare-IoT-
    security-issues- Risks-and-what-to-do-about-them.

25. U.S. Department of Health and Human Issues. "Summary of the HIPPA Security Rule."
    Retrieved from https://www.hhs.gov/hipaa/for-professionals/security/laws-regulations/
    index.html.

11 Deep Learning for the Prediction and Detection of Alzheimer's Disease (AD)

An Overview and Future Trends

Pavanalaxmi, Roopashree

CONTENTS

11.1 INTRODUCTION

Alzheimer's disease (AD), the most well-known category of dementia, is an unpredictable illness portrayed by a gathering of β-amyloid (Aβ) plaques and neurofibrillary tangles made out of tau amyloid fibrils [1] related through neurotransmitter misfortune and neurodegeneration, prompting memory impedance and extra intellectual issues.

In worlds population 50 million people are having AD related symptoms as per Bright Focus Foundation [2]. As per the 2019 World Alzheimer report [2], in the world, around 50 million people have AD or related dementia. For every four individuals, one person is affected by AD. In Western Europe AD is commonly seen but in the case of Sub-Saharan Africa frequency of AD is very less compared to Western Europe as per the data available in [2]. It's predicted that by 2050, 68% of the growth in AD worldwide will be seen in lower and middle socioeconomic countries[X].

In 2018, AD and dementia care in the United States cost $277 billion USD. In 2018, family members of individuals with AD or other dementias gave an expected 18.5 billion hours of care, which is worth an estimated at $233.9 billion USD in unpaid time. The complete lifetime cost of dementia patient price of care for a dementia patient was assessed at $250,174 USD in 2018. By 2050, expenses related to dementia could be $1.1 trillion USD [2].

The principal manifestations of AD fluctuates in individuals. Memory difficulties are regularly one of the important signs of mental impedance recognized with AD. Also, some patients might be determined to have mild intellectual impedance. As the illness spreads, the patient experiences more noteworthy cognitive decline and extra psychological troubles.

AD progresses in limited stages: premedical, gentle (also known as beginning phase), moderate, and extreme (in some cases known as late-stage). In gentle AD, an individual may seem, by all accounts, to be well but will encounters progressively more trouble understanding his overall environmental factors. The feeling that something isn't right consistently comes gradually to the patient and their family. In a moderate stage, more heightened oversight and care become needed, which can be hard for partners and family.

11.1.1 HAZARD COMPONENTS OF ALZHEIMER'S DISEASE

While the specific cause of AD isn't clear, certain hazard components are unmistakably connected with the advancement of AD. The next sections discuss the entirety of hazard components in more depth in their appropriate contexts.

- **Age**: Age is a factor that is associated with AD. After the age of 65, the threat of AD duplicates predictably and is at about a 50% risk after the age of 85 [2].
- **Family ancestry**: Close relatives to AD patients are likely to develop it as well. If more than one relative has the illness, the likelihood increases. When diseases tend to run in families, hereditary and environmental factors will play a major role [2].
- **Genetics**: Genetics may moreover predict a huge part in the growth of AD. There are two classes of characteristics that impact Alzheimer's change: risk

genes (the growth in probability) and deterministic genes (genes that are the cause of an ailment).

11.1.2 SYMPTOMS

Indications of AD regularly start with trouble in recalling recently learned data, prompting progressively serious indications, for example, "disorientation, mood, and behavior changes; deepening confusion about events, time and place; unfounded suspicions about family, friends and professional caregivers; more serious memory loss and behavior changes; and difficulty speaking, swallowing and walking" [3] as it progresses. Side effects are more noticeable by family or companions than to the influenced person.

The hippocampus and the entorhinal cortex will be damaged during the initial stage of AD. The brain tissue will shrink as more neurons die. During the last phase of AD, the cerebrum tissue will essentially shrivel. Harmful changes are occurring in the brain during this preclinical phase of AD. At the point when sound neurons stop working brain will lose the associations with different neurons. At that point, in the back of the cerebrum, proteins begin to eradicate the psyche's most profound recollections. Finally, control focuses (overseeing heart rate and breathing) are overwhelmed, causing death.

11.1.3 DIAGNOSIS

New symptomatic standards were given for Mild Cognitive Impairment because of AD, just as Dementia standards were updated because of AD in 2011. This was a result of the understanding in the medical field that the malady joins a certain stage when impacted individuals encounter a moderate increase of learning capacities that result from the assortment of AD pathology in the brain. The patient is determined to have AD dementia at the point when side effects are adequately articulated (for example with extraordinary psychological impedance). The scholars of the models noticed "as AD is a moderate, reformist issue, with no fixed occasions that characterize its beginning, it is especially trying for clinicians to distinguish change focuses for singular patients" [4], and the sufferer changes from the asymptomatic stage to the suggestive predementia stage. The suggestive predementia stage to the stage that marks the beginning of dementia is also hard to distinguish.

11.1.4 MILD COGNITIVE IMPAIRMENT BECAUSE OF ALZHEIMER'S DISEASE

Mild cognitive impairment (MCI) due to AD is a syndrome characterized by "clinical, cognitive, and practical standards" [4] and is the suggested predementia stage of Alzheimer's ailment. The investigative norms are:

- Concern for an adjustment in reasoning
- Damage in at least one cognitive space
- Protection of autonomy in effective capacities

On the off chance that a patient meets these standards, the clinical specialist needs to evaluate whether there is target proof of psychological decay from reports by the patient, and additionally different witnesses. Since other psychological areas can be disabled in MCI patients, clinicians must likewise analyze different spaces than memory, for example, chief capacities, language, visuospatial abilities, and attention control. There are various approved medical neuropsychological examinations available for this reason. If these tests are not possible, there are likewise assortments of basic casual methods that can be utilized. Researchers likewise bring up that "a typical clinical introduction of AD may emerge, such as the visual variation of AD or the language variation (at times called logopenic aphasia), and these clinical profiles are additionally steady with MCI because of AD" [4].

11.1.5 DEMENTIA BECAUSE OF ALZHEIMER'S DISEASE

The changed measures for AD allowed healthcare staff to treat patients without first admitting them to a neuropsychological unit, and helped in examinations through progressed imaging, cerebrospinal liquid measures, and connection to specialists associated with research or in clinical preliminary investigations that might have these instruments accessible [5]. Patients are analyzed when there are some of these psychological side effects:

a. Change in the capacity to function at work or common happenings
b. Speech regresses compared to past performance
c. Symptoms are not explained by insanity or other mental issue diagnosis
d. Cognitive weakness is identified and analyzed using a mixture of
   • Age of a patient and an witness who knows them well and
   • A target cognitive appraisal or a "bedside" psychological position assessment
e. The psychological or conduct weakness includes at least two of the associated parts:
   • Delayed ability to gain and remember fresh data
   • Debilitated thinking and managing of complex work, low judgment
   • Weakened visuospatial capacities
   • Weakened language capacities
   • Changes in character, conduct, or comportment

11.1.6 TREATMENT

Right now, there is no cure for AD. The aim of medicine is to reduce the progression of the ailment, which means controlling its side effects. It is conceivable to a limited degree to manage AD well on the off chance that it is analyzed generally at an early stage. Medicines oversee side effects, for example, intellectual, mental issues, and conduct issues, and offer the natural acclimation to empower subjects to more readily perform everyday exercises, aiding guardians and relatives.

11.2 ORGANIZATION OF THIS CHAPTER

This chapter has an outline of various deep learning-based methods used for AD diagnosis. The main aim of this chapter is:

1. In Section 11.1 we discuss Alzheimer's disease.
2. In Section 11.3, the deep learning-based structure is presented and also organizations are summarized.
3. Sections 11.4 and 11.5 describe different types of image management techniques used for AD detection and we have reviewed some of the AD-related papers.

11.3 DEEP NEURAL NETWORKS

11.3.1 GENERAL ARCHITECTURE OF DEEP NEURAL NETWORKS

Different profound structures were developed by customary feed-forward artificial neural networks (ANNs). An ANN comprises multi-stage layers. The input and output of every layer have a set of arrays known as feature maps. Every feature map in a particular layer speaks to specific highlights removed in the areas of the related information.

Different layers in deep learning (DL) are mentioned below:

- **Input layer:** The convolutional layer takes the input from this layer. A few changes, for example, mean-deduction, highlight scaling, and compelling information enlargement can be incorporated [6].
- **Convolutional layer**: incorporates three phases of operational elements [7]
  - **Convolutional filters**: These calculate the convolution result of the input feature map with trainable two-dimensional discrete convolution filters and bias parameters. Each filter bank distinguishes a specific element for every area on the input map. [8]
  - **Pooling**: The spatial component of the input is undersampled using the pooling technique. This will decrease the results of the output feature map in terms of the features in the previous layer. Various varieties of pooling are available [9].
  - **Nonlinearity function**: This is a nonlinear element-wise operator used to simulate the excitability of neurons. Image processing uses ReLU nonlinearity function in DL.
- **Normalization layer**: This executes at each spatial area overall feature maps of a similar layer for obtaining the improvement in the input [10].
- **Dropout regularization layer**: Overfitting of the network can be reduced and also more features can be learned by using this layer. The key thought is to arbitrarily drop alongside the separate associations from the neural organization during the preparation cycle to evade an excessive amount of covariation of the units [11].
- **Fully connected layers**: The input as well as the output is of vector form. The final layers act as fully connected layers so that it can be used for the classification [3].

Using these components, the signals can be made propagated from layer to layer using active neurons in the architecture designed.

11.3.2 CONVOLUTIONAL NEURAL NETWORKS

Picture recognition utilizes the feed-forward ANN. The convolution neural network (CNN) structure is the subject of significant enthusiasm for computerized image preparation and vision. Without any preprocessing, we were able to extract the features and patterns that were learned by the researchers. Feature extractors and trainable classifiers are the two layers used by the CNN. VGGNet, googleNEt, Alexnet, faster R-CNN, ResNet, and ZFnet are some of the various kinds of CNN architecture. Correlation of various structures of profound learning is shown in Table 11.1.

11.4 INPUT DATA MANAGEMENT

11.4.1 BIOLOGICAL MARKERS AND CHARACTERISTICS OF A PHENOMENON BEING OBSERVED IN AD RECOGNITION

To precisely identify AD toward the early phases of sickness requires an assessment of some quantitative biological markers. Magnetic resonance imaging (MRI) is the most generally accessible as well as most utilized biological marker for AD location. It creates a three-dimensional (3-D) portrayal of various regions of the body by using a dominant magnetic field and radio-frequency (RF) pulses. fMRI imitates the progressions related to the bloodstream. Positron emission tomography (PET) scanning depends on nuclear medication strategies that can perceive metabolic cycles inside the body. We can consider many factors other than neuroimaging techniques to detect AD: age, sex, instructive level, coordination and balance, sense of sight and hearing, electroencephalogram, cerebrospinal fluid (CSF) biomarkers, and psychological function known to be linked to a particular brain structure. These components, alongside the different neuroimaging modalities, can entangle the preparation of DL models.

Neuroimaging data can be used to extract many features from the image. Feature extraction methods are used to make a lot of precise data. The extracted information should indicate the disease's design and be promptly characterized. All cataloging algorithms have three phases: feature extraction, feature dimension reduction, and characterization. Using DL methods, these three phases can be combined into one. Yet, managing the whole neuroimaging philosophy is so far a test. The methodologies to input information organization can generally be assembled into four classes [12, 13].

1. **Voxel based**: This method uses voxel concentration standards from whole neuroimaging modalities. Prior to deep model application, this approach can be used to analyze either full brain images or on dimensional detailed anatomical images. A voxel preselection technique is suggested to utilize to best large dimension of the feature.
2. **Slice based**: Two-dimensional (2-D) image slices can be abstracted from a 3-D brain scan using unique techniques available. Sagittal, coronal, and

TABLE 11.1

Correlation of Various Structures of Profound Learning

Types of Structure	Detail of Structures	Advantages	Disadvantages
Deep neural network (DNN)	There are two or more layers, which permits a complex nonlinear state. It is utilized by grouping and regression.	It is broadly utilized with incredible accuracy.	In the training process, erroneous data are transferred back to the earlier single layers. The learning cycle of the model is moderate.
Convolutional neural network (CNN)	It is generally excellent for bidirectional information. It comprises convolutional channels that change two dimensions (2-D) to three dimensions (3-D).	Generally excellent execution, model adapts quickly.	It needs a lot of marked information for grouping.
Recurrent neural network	It can learn temporal arrangement. The weights are common overall path and neurons.	These give cutting edge precision in voice detection, character detection, and numerous other Natural language processing related undertakings.	Here numerous issues because of a disappearing slope and the requirement for enormous database.
Deep Conventional Extreme Learning Machine	This organization utilizes a Gaussian likelihood work for the sampling of local connections.	It is a proficient process. It has a quick training execution. It is useful for random impairment.	Formatting can be successful if the acquisition work is straightforward but also if the measure of tagged information is minimum.
Deep Boltzmann Machine	It depends on the group of Boltzmann devices. The model comprises simplex associations with completely invisible layers.	The top-down input consolidates uncertain information for much robust conclusion.	Advancement of factor isn't workable for an enormous database.
Deep belief network	This uses one directional relation and is utilized in supervised as well as in unsupervised machine learning (ML). The invisible layers of each subnetwork help as a seeable layer for the succeeding layer.	The methodology (utilized in each layer) and the induction manageable boost legitimately the probability.	The formatting makes the preparation procedure computationally exclusive.
Deep Autoencoder	This structure is utilized in unsupervised learning. It is used mostly to extract and reduce the features. The input and output information is of the same amount.	Labeled data isn't required.	This requires a preprocessing stage.

axial plane projection methods also can convert a 3-D image to obtain 2-D image slices. These techniques will not perform complete brain investigation, as a 2-D image slice excludes some information from a brain scan. These techniques commonly document the central nervous system and disregard the remaining parts.

3. **ROI based**: Region of Interest (ROI) techniques emphasize specific portions of the brain recognized to be influenced in the beginning phases of AD. ROI for the most part requires past information of the irregular regions and a cerebrum chart joined with the long-term involvement of scientists.

4. **Patch based**: These methodologies can catch infection associated configurations in a cerebrum by taking out structures from patches. This technique collects useful patches, which are used to extract both patch- and image-level features [14]. This method is utilized in numerous examinations for AD identification [15].

Handling the input information to deep models is also a significant concern. Three-dimensional images will generate large training parameters, so using 2-D slices as input avoids producing a large number of training parameters and results in improved organizations. Voxel-based strategies can fit all 3-D data in a single brain scan. So, voxel preselection techniques could be used. It is difficult to perform classification and to extract features from a distinct brain scan as it comprises a composite configuration of voxels and a lot of information. It is important to extract a set quantity of distinct predefined representative regions. The ROI-based approaches are effortlessly deduced and executed in medical preparation. The dimensionality of ROI-based features is continually slighter as the complete brain is characterized by smaller features. Patch-based procedures can powerfully deal with high feature dimensions and produce little deviation. Since patch abstraction doesn't need ROI identification, the need to include a human professional is decreased contrasted with ROI-based methodologies.

11.5   CASE STUDIES

Some of the case studies are described in this section.

Hosseini-Asl et al. [16] proposed a 3-D Adaptable CNN (3-D A-CNN) classifier to estimate AD on structural brain MRI scans. It is another deep 3-D CNN utilized for extracting features, which is dependent on a 3-D convolutional Autoencoder. Masci has proposed a 3-D convolutional Autoencoder [17]. A mass of unsupervised convolutional autoencoder (CAE) is utilized to extract local features from 3-D images. Each information picture is condensed hierarchically for training the following layer CAE. To catch characteristic dissimilarities of an information 3-D image, each voxel-wise feature is separated using a linear encoding filter. The lower layers extract universal features and the higher layers need to encourage task specific organization [18]. The classifier extracts the universal features by utilizing a stack of privately associated bottom convolutional layers and completely associated upper layers. At the initial phase, the convolutional layers need to separate generic features, which are framed as a stack of 3-D CAEs and are pretrained. These layers are encoded by 3-D CAE weights and the fine tuning of the upper layer is performed for structural MRI arrangement. The base convolutional layers can separate common features related to the disease biological markers. The proposed execution of the 3-D CNN utilizes the ReLU activation utilities at each inner layer and the completely associated upper layers with a softmax top most output layer to categorize an input brain image. The *CAD Dementia* Dataset was used to pretrain the 3-D CAE network. Next the learned features are extracted and utilized as AD biological marker detection in the lowest layers of a 3-D CNN network. For AD grouping on 210 subjects, the ADNI data set is used in the experiment performed.

Hu et al. [19] used AD as a reference to demonstrate the benefits of DL in analyzing brain diseases and providing medical result support. The method diminishes the calculus complication and articulates the links amongst different areas of the cerebrum. An adapted autoencoder design is constructed for the arrangement of patients and AD estimation. The architecture would extract features of the entire cerebrum, inspect the correspondence of different regional functions, and harvest the whole assessment of the brain's intellectual ability. The precision of disease estimate claims to make 25% more enhancement than traditional methods. The training samples are chosen from the ADNI database. The fMRI information downloaded is in DICOM format originally. The DPABI toolbox [20] is utilized to handle this information. The signals are made steadier by eliminating pictures that are in the time domain. Then, preprocessing is done. This procedure includes the subdivision of the time domain, rearrangement, uniformization, and smoothing. Finally, the Automated Anatomical Labeling (AAL) model is utilized to recognize the cerebrum ROI. This partitions the entire cerebrum into 90 areas. A correlation matrix is acquired, by ascertaining the relationship among different areas of cerebrum. A focused auto encoder (AE) system is worked to order the correlation matrix, which is sensitive to AD. The time-series value and the correlation-coefficient value have been placed into various machine learning (ML) models to train individually. Support-vector machines, Logistic Regression (LR), and the AE models were utilized to learn and examine brain features. Every machine learning model is trained 10 times and 10 dissimilar grouping outcomes are obtained. The mean value of the 10 accuracies is computed and the integral performance of the ML model is characterized. The analysis outcomes show that the proposed technique for AD prediction accomplishes considerable improved effects. This discovers the associations among various cerebrum areas professionally and gives a solid reference for AD estimation. It helps to predict AD at the beginning phase and bring methods to reduce or even avoid the onset of it.

Vu et al. [21] proposed a DL technique on blended multimodalities. The suggested methodology combines Sparse Autoencoder (SAE) and CNN training and testing is done on united PET-MRI information to identify the infection status of a patient. The classification accuracy is improved by including the multimodalities rather than just one. The suggested technique can accomplish a grouping accurateness of 90% between AD patients and normal controls. An autoencoder is trained on a lot of arbitrarily chosen 3-D patches taken out from the MRI scans and same number of patches from the PET scans. Some sets of scans were used to train, validate, and test. Thus, each autoencoder is trained with 40,000 patches, then fine-tuned with 5,000 authenticated patches and verified with remaining patches.

After training the sparse-AE, a 3-D CNN is built, which considers MRI and a PET scan together. In this work, max-pooling layers are used followed by convolutional layers. The outputs are then arranged and utilized as inputs for a three-layer completely associated NN. A sigmoid function for the hidden layer and softmax activation function for an output layer were set. The result obtained represents the conditional probabilities that showed whether the information has a place with AD or normal control (NC) class. The ADNI dataset is used for the experiment, which is publicly available on the internet. Precisely, MRI and PET data attained from 145 AD matters and 172 NC matters were deliberated. Both the images were preprocessed using different preprocessing procedures. By coregistering the PET pictures to their individual MRI images, they were spatially standardized. Statistical parametric mapping was utilized to standardize for brain mapping template. The data is normalized by subtracting the mean and dividing by the standard deviation. The early-stopping time is determined using a validation set of 127. The enactment of the model was assessed using a test sample of 191 examples. The fusion method shows better prospective in capturing local 3-D patterns from both these imaging techniques, which is useful in the estimation of many brain-related infections that can be interpreted through the test results. The suggested technique improves the classification accuracy.

Oh et al. [22] present a methodology intended to urge the end-to-end learning of a volumetric CNN model for organization tasks dependent on MRI and pictures its results without any human intervention. DL models permit a structure to utilize fresh data as input, which allows finding unique features from the given dataset [23]. MRI scans were pretrained dependent on CAE-based unsupervised learning. The ADNI dataset has been used for the proposed system. Supervised fine-tuning was directed to construct the classifier to distinguish AD from normal controls. The supervised transfer learning method is used to resolve other classification tasks. The visual portrayals drawn from the grouping assignment of AD/normal controls were moved to the pMCI/sMCI learning model. At last, the most significant biological markers on each grouping task were identified utilizing the saliency visualization method. Here, 694 sMRI scans that were primarily categorized into AD ($n = 198$), normal controls ($n = 230$), progressive MCI ($n = 166$), and stable MCI ($n = 101$) at baseline were utilized for classification. Because of the restricted measure of information, data augmentation was performed. The proposed approach utilizes CAE based unsupervised learning, and supervised transfer learning to solve different organization task. A gradient-based visualization technique was applied for identifying the most relevant biological markers associated with AD and pMCI. To authenticate the offerings of this learning, trials were conducted on the ADNI database and it attained better accuracy for classification task.

Jyoti [24] delivered a unique DL model for analyzing brain MRI scans and categorizing them into diverse AD phases. The suggested deep CNN model can identify early phases of AD and categorize them. A very deep CN is designed and the performance on the OASIS database is demonstrated. The proposed deep CNN model for AD recognition and organization used Tensorflow and Python. The proposed model is claimed to be much faster than all previous traditional methods taking less than 1 hour to train and test for AD discovery and organization.

Lin et al. [25] has utilized data, which were considered from the ADNI database, which included 188 AD, 229 normal controls, and 401 MCI subjects. In this examination, a DL approach established on CNN on MRI data is aimed at precisely prediction of MCI-to-AD conversion. Age-correction and processing techniques are used on MRI data. Local patches are extracted from these images. CNN-based architecture is constructed to extract high-level features for classification. The patches from AD and normal controls are utilized to train CNNs to distinguish DL features of MCI subjects. The prediction is improved by adding more morphological information. Structural MRI image features are mined with FreeSurfer to assist the CNN. Finally, both CNN and FreeSurfer-based features are fed into an extreme learning machine classifier to guess the AD conversion. Validation of the suggested method was done on the standardized MRI datasets from the ADNI database. This method claims to attain better accuracy compared to all other available models. The suggested approach outperforms other state-of-the-art approaches available with higher accuracy and AUC. Results show that with only MRI data, it is easy to predict the MCI-to-AD conversion using the proposed CNN-based method. If the age-correction method and assisted structural brain image features are used, estimate performance of CNN can be improved.

Jyoti Islam and Yanqing Zhang [26] established a deep CNN that extracts features from the input sMRI images. They have trained the model using the OASIS database and the proposed model can recognize AD and classify the current disease phase. The model categorized the four phases of the AD. The input MRI is 3-D data, and the suggested method is a 2-D architecture, so the input data is converted into 2-D images initially. From three physical planes of imaging, patches were created for each set of MRI data. The patches created are axial, coronal, and sagittal planes. The proposed network utilizes these patches as input. The number of samples in the training dataset is improved by means of this data augmentation procedure. Each model has its own softmax layer and they are trained separately. The softmax layer classifies the input image into four different AD stages based on feature representation by individual model. The proposed network claimed to be advantageous for early-stage AD identification.

Classification of a model for AD vs. MCI vs. NC groupings can be constructed by carrying out transfer learning (TL) using CNN architecture, VGG16 [27]. Completely associated layers are added on top of VGG16, which extracts features. The layers of the base model are kept nontrainable and the resultant mathematical model acquired is trained on brain MRI slices. By comparing the entropy, the overall strength of the model is improved by using the sMRI slices of subjects with the most relevant information. The T1-weighted sMRI data selected 50 AD, 50 cognitively normal, and 50 MCI cases for the classification task from the ADNI database. Initially, cortical reconstruction and FreeSurfer suite volumetric segmentation were performed to eliminate unnecessary details. After preprocessing brain MRI images, the resultant image comprises 2-D images called slices. There are 256 slices corresponding to 3-D images. Using an image entropy-based sorting mechanism, the most informative slices of an image is made and only 32 slices were chosen of each subject. VGG16 was trained on RGB. VGG16 is pretrained on ImageNet dataset to extract the feature values having a place with that specific MRI slice. The model for three-way organization was tailored on training information. The proposed organization model takes about 10 hours for training the set. Accuracy is calculated, which is used as the primary evaluation metric. The resultant accurateness for the validation set obtained is 95.73% for the proposed approach. For binary classifications, models were able to attain the highest accuracy for AD vs. cognitively normal, AD vs. MCI, and MCI vs. cognitively normal classifications, respectively. Different metrics were processed, which supports the presentation of the suggested classification model and the exactness of the method is compared with past methods.

11.6 COMPARATIVE ANALYSIS

The source code for most of the proposed models is not accessible online or offline. Therefore, only a partial comparison of these models can be done, but may not give proper information. Major studies have just reported their final accuracies instead of actually implementing tem. This section gives comparative information collected from different review papers considered. In most of the studies, training of DL is done from scratch, but it is inefficient as the training process is time-consuming and millions of images have to be considered. Hence, it is better to use some techniques to reduce the size of the dataset or generate a neuroimaging dataset that has only hundreds of images. Using transfer learning methods, neuroimaging datasets can be created more quickly and it accomplishes better outcomes as compared to training from scratch [28–30].

A few notable 2-D CNNs utilizes transfer learning method [31]. A blend of patch-based and ROI-based approaches [32] gave higher exactness than a ROI-based technique [33] that utilized stacked autoencoders. A blend of patch-based and ROI-based with stacked autoencoders [34] outflanked a mix of patch-based and voxel-based Deep Boltzmann Machine [35]. Voxel-based 3-D CNN and stacked 3-D autoencoders [16] were accounted for to be more exact contrasted and ROI-based stacked autoencoders [33, 36] and Restricted Boltzmann Machine (RBM)s [37]. Slice centered VGGNet-16 [38] beat a patch-centered [15] and voxel-centered [39] mix of autoencoders and CNNs. Concerning input information organization approaches, ROI and patch-centered techniques are more effective than related methods. Among DL models, a stacked autoencoder can fundamentally increase the illustrative intensity of highly nonlinear and complex patterns. Autoencoders can discover good initialization parameters for CNNs, which is known to be beneficial. Supervised methods were described [40] to have improved presentation related with autoencoders when a deep polynomial network or a stacked deep polynomial network is used. DNNs are appropriate to vector-based issues when related to support vector machines, and the learning process is excessively moderate [41]. Three-dimensional CNNs and 2-D CNNs are the most popular supervised methods, which are enhanced for image-related issues. The 3-D CNNs can catch 3-D data from the 3-D volume of a brain scan. Three-dimensional CNNs have indicated better execution compared with 2-D CNNs [42], but training is more complex, which can be settled utilizing patch-based or ROI-based strategies instead of voxel-based ones. Two-dimensional CNNs are simpler to train, however a plan utilizing it isn't effective in encoding the spatial data of the 3-D images. Therefore, RNNs are mostly used after 2-D CNN to capture 3-D data in adjacent image slices in a sequence of images. There are many cerebrum image investigation bundles such as FreeSurfer9, FSL 10, MIPAV 11, and SPM 12 that offer powerful tools to various automated preprocessing methods [43]. Deep models can be implemented using software bundles such as MATLAB 13, Keras 14, Tensorflow 15, Theano 16, Caffe 17, and Torch 18 [44]. In addition to these, there are online datasets such as ADNI [45], AIBL [46], OASIS [47], and MIRIAD 22 [48] that help to simplify AD detection task.

11.7   CONCLUSION

Alzheimer's disease is the most common sources of death, particularly in urbanized countries. It is a difficult task to identify AD in its initial phases, the utilization of computer centered frameworks along with clinical specialists has a lot to offer in identifying AD. DL can be used as a powerful method in the early recognition of AD. This chapter is a review of some of the approaches employed for improving AD discovery based on DL and neuroimaging modalities. This chapter has a brief overview of AD and different DL structures that could be employed for early detection of AD. We have discussed the different inputs to be used for detection. A few case studies were also reviewed in this study along with comparative analysis.

REFERENCES

1. Hardy, John. "Alzheimer's disease: The amyloid cascade hypothesis: An update and reappraisal." *Journal of Alzheimer's Disease* 9, no. s3(2006): 151–153.
2. Alzheimer's Association (website). "Alzheimer's & Dementia Causes, Risk Factors." https://www.alzheimers.net/alzheimers-statistics.
3. Krizhevsky, Alex, Ilya Sutskever, and Geoffrey E. Hinton. "Imagenet classification with deep convolutional neural networks." *Advances in Neural Information Processing Systems* 25 (2012): 1097–1105.
4. Albert, Marilyn S., Steven T. DeKosky, Dennis Dickson, Bruno Dubois, Howard H. Feldman, Nick C. Fox, Anthony Gamst, et al. "The diagnosis of mild cognitive impairment due to Alzheimer's disease: Recommendations from the National Institute on Aging—Alzheimer's Association workgroups on diagnostic guidelines for Alzheimer's disease." *Alzheimer's and Dementia* 7, no. 3 (2011): 270–279.
5. McKhann, Guy M., David S. Knopman, Howard Chertkow, Bradley T. Hyman, Clifford R. Jack Jr, Claudia H. Kawas, William E. Klunk, et al. "The diagnosis of dementia due to Alzheimer's disease: recommendations from the National Institute on Aging-Alzheimer's Association workgroups on diagnostic guidelines for Alzheimer's disease." *Alzheimer's and Dementia* 7, no. 3 (2011): 263–269.
6. Hamidinekoo, Azam., Zobia Suhail, Tallha Qaiser, Reyer Zwiggelaar. "Investigating the effect of various augmentations on the input data fed to a convolutional neural network for the task of mammographic mass classification." In: Valdés Hernández M., González-Castro V. (Eds.), In Annual Conference on Medical Image Understanding and Analysis. MIUA 2017. Communications in Computer and Information Science, vol 723. Springer: Cham, 2017.
7. LeCun, Yann, Koray Kavukcuoglu, and Clement. Farabet. "Convolutional networks and applications in vision," In: Proceedings of 2010 IEEE International Symposium on Circuits and Systems, Paris, 2010, pp. 253–256.

8. Schmidhuber, Jürgen. "Deep learning in neural networks: An overview." *Neural Networks* 61 (2015): 85–117.
9. Krizhevsky, Alex, and Geoffrey Hinton. "Learning multiple layers of features from tiny images." 2009. http://citeseerx.ist.psu.edu/viewdoc/download?doi=10.1.1.222.9220&rep=rep1&type=pdf
10. Dahl, George E., Tara N. Sainath, and Geoffrey E. Hinton. "Improving deep neural networks for LVCSR using rectified linear units and dropout." In 2013 IEEE international conference on acoustics, speech and signal processing, pp. 8609–8613. IEEE, 2013.
11. Srivastava, Nitish, Geoffrey Hinton, Alex Krizhevsky, Ilya Sutskever, and Ruslan Salakhutdinov. "Dropout: A simple way to prevent neural networks from overfitting." *The Journal of Machine Learning Research* 15, no. 1 (2014): 1929–1958.
12. Zheng, Chuanchuan, Yong Xia, Yongsheng Pan, and Jinhu Chen. "Automated identification of dementia using medical imaging: a survey from a pattern classification perspective." *Brain Informatics* 3, no. 1 (2016): 17–27.
13. Cuingnet, Rémi, Emilie Gerardin, Jérôme Tessieras, Guillaume Auzias, Stéphane Lehéricy, Marie-Odile Habert, Marie Chupin, Habib Benali, Olivier Colliot, and Alzheimer's Disease Neuroimaging Initiative. "Automatic classification of patients with Alzheimer's disease from structural MRI: A comparison of ten methods using the ADNI database." *NeuroImage* 56, no. 2 (2011): 766–781.
14. Liu, Mingxia, Jun Zhang, Ehsan Adeli, and Dinggang Shen. "Landmark-based deep multi-instance learning for brain disease diagnosis." *Medical Image Analysis* 43 (2018): 157–168.
15. Gupta, Ashish, Murat Ayhan, and Anthony Maida. "Natural image bases to represent neuroimaging data." In Proceedings of the 30th International Conference on Machine Learning, pp. 987–994. 2013.
16. Hosseini-Asl, Ehsan, Robert Keynton, and Ayman El-Baz. "Alzheimer's disease diagnostics by adaptation of 3D convolutional network." In 2016 IEEE International Conference on Image Processing (ICIP), pp. 126–130. IEEE, 2016.
17. Masci, Jonathan, Ueli Meier, Dan Cireşan, and Jürgen Schmidhuber. "Stacked convolutional auto-encoders for hierarchical feature extraction." In ICANN 2011, International Conference on Artificial Neural Networks, pp. 52–59. Springer, Berlin, Heidelberg, 2011.
18. Long, Mingsheng, Yue Cao, Jianmin Wang, and Michael Jordan. "Learning transferable features with deep adaptation networks." In Proceedings of the 32nd International Conference on Machine Learning, pp. 97–105. PMLR, 2015.
19. Hu, Chenhui, Ronghui Ju, Yusong Shen, Pan Zhou, and Quanzheng Li. "Clinical decision support for Alzheimer's disease based on deep learning and brain network." In 2016 IEEE International Conference on Communications (ICC), pp. 1–6. IEEE, 2016.
20. Chao-Gan, Y. A. N. "DPABI: A toolbox for Data Processing & Analysis for Brain Imaging." (2014). http://rfmri.org/dpabi
21. Vu, Tien Duong, Hyung-Jeong Yang, Van Quan Nguyen, A-Ran Oh, and Mi-Sun Kim. "Multimodal learning using convolution neural network and Sparse Autoencoder." In 2017 IEEE International Conference on Big Data and Smart Computing (BigComp), pp. 309–312. IEEE, 2017.
22. Oh, Kanghan, Young-Chul Chung, Ko Woon Kim, Woo-Sung Kim, and Il-Seok Oh. "Classification and visualization of Alzheimer's disease using volumetric convolutional neural network and transfer learning." *Scientific Reports* 9, no. 1 (2019): 1–16.
23. Shen, Dinggang, Guorong Wu, and Heung-Il Suk. "Deep learning in medical image analysis." *Annual Review of Biomedical Engineering* 19 (2017): 221–248.
24. Islam, Jyoti, and Yanqing Zhang. "A novel deep learning based multi-class classification method for Alzheimer's disease detection using brain MRI data." In International Conference on Brain Informatics, pp. 213–222. Springer, Cham, 2017.

25. Lin, Weiming, Tong Tong, Qinquan Gao, Di Guo, Xiaofeng Du, Yonggui Yang, Gang Guo, et al. "Convolutional neural networks-based MRI image analysis for the Alzheimer's disease prediction from mild cognitive impairment." *Frontiers in Neuroscience* 12 (2018): 777.
26. Islam, Jyoti, and Yanqing Zhang. "Brain MRI analysis for Alzheimer's disease diagnosis using an ensemble system of deep convolutional neural networks." *Brain Informatics* 5, no. 2 (2018): 2.
27. Jain, Rachna, Nikita Jain, Akshay Aggarwal, and D. Jude Hemanth. "Convolutional neural network based Alzheimer's disease classification from magnetic resonance brain images." *Cognitive Systems Research* 57 (2019): 147–159.
28. Valliani, Aly, and Ameet Soni. "Deep residual nets for improved Alzheimer's diagnosis." In Proceedings of the 8th ACM International Conference on Bioinformatics, Computational Biology, and Health Informatics, pp. 615–615. 2017.
29. Wegmayr, V. and Haziza, D., 2018. "Alzheimer Classification with MR images: Exploration of CNN Performance Factors" 1st Conference on Medical Imaging with Deep Learning (MIDL 2018), Amsterdam, The Netherlands.
30. Hon, Marcia, and Naimul Mefraz Khan. "Towards Alzheimer's disease classification through transfer learning." In 2017 IEEE International Conference on Bioinformatics and Biomedicine (BIBM), pp. 1166–1169. IEEE, 2017.
31. Ebrahimi-Ghahnavieh, Amir, Suhuai Luo, and Raymond Chiong. "Transfer learning for Alzheimer's disease detection on MRI images." In 2019 IEEE International Conference on Industry 4.0, Artificial Intelligence, and Communications Technology (IAICT), pp. 133–138. IEEE, 2019.
32. Shi, Bibo, Yani Chen, Pin Zhang, Charles D. Smith, Jundong Liu, and Alzheimer's Disease Neuroimaging Initiative. "Nonlinear feature transformation and deep fusion for Alzheimer's Disease staging analysis." *Pattern Recognition* 63 (2017): 487–498.
33. Liu, Siqi, Sidong Liu, Weidong Cai, Hangyu Che, Sonia Pujol, Ron Kikinis, Dagan Feng, and Michael J. Fulham. "Multimodal neuroimaging feature learning for multiclass diagnosis of Alzheimer's disease." *IEEE Transactions on Biomedical Engineering* 62, no. 4 (2014): 1132–1140.
34. Lu, Donghuan, Karteek Popuri, Gavin Weiguang Ding, Rakesh Balachandar, and Mirza Faisal Beg. "Multimodal and multiscale deep neural networks for the early diagnosis of Alzheimer's disease using structural MR and FDG-PET images." *Scientific Reports* 8, no. 1 (2018): 1–13.
35. Suk, Heung-Il, Seong-Whan Lee, Dinggang Shen, and Alzheimer's Disease Neuroimaging Initiative. "Hierarchical feature representation and multimodal fusion with deep learning for AD/MCI diagnosis." *NeuroImage* 101 (2014): 569–582.
36. Suk, Heung-Il, and Dinggang Shen. "Deep learning-based feature representation for AD/MCI classification." In International Conference on Medical Image Computing and Computer-Assisted Intervention, pp. 583–590. Springer, Berlin, Heidelberg, 2013.
37. Li, Feng, Loc Tran, Kim-Han Thung, Shuiwang Ji, Dinggang Shen, and Jiang Li. "A robust deep model for improved classification of AD/MCI patients." *IEEE Journal of Biomedical and Health Informatics* 19, no. 5 (2015): 1610–1616.
38. Billones, Ciprian D., Olivia Jan Louville D. Demetria, David Earl D. Hostallero, and Prospero C. Naval. "DemNet: A convolutional neural network for the detection of Alzheimer's disease and mild cognitive impairment." In 2016 IEEE Region 10 Conference (TENCON), pp. 3724–3727. IEEE, 2016.
39. Payan, Adrien, and Giovanni Montana. "Predicting Alzheimer's disease: A neuroimaging study with 3D convolutional neural networks." arXiv preprint arXiv:1502.02506 (2015).
40. Zheng, Xiao, Jun Shi, Yan Li, Xiao Liu, and Qi Zhang. "Multi-modality stacked deep polynomial network based feature learning for Alzheimer's disease diagnosis." In 2016

IEEE 13th International Symposium on Biomedical Imaging (ISBI), pp. 851–854. IEEE, 2016.

41. Razzak, Muhammad Imran, Saeeda Naz, and Ahmad Zaib. "Deep learning for medical image processing: Overview, challenges and the future." In Classification in BioApps, pp. 323–350. Springer, Cham, 2018.

42. Tang, Hao, Erlin Yao, Guangming Tan, and Xiuhua Guo. "A Fast and Accurate 3D Fine- Tuning Convolutional Neural Network for Alzheimer's Disease Diagnosis." In International CCF Conference on Artificial Intelligence, pp. 115–126. Springer, Singapore, 2018.

43. Vinutha, N., P. Deepa Shenoy, and K. R. Venugopal. "Efficient Morphometric Techniques in Alzheimer's Disease Detection: Survey and Tools." *Neuroscience International* 7, no. 2 (2016): 19–44.

44. Rouast, Philipp V., Marc Adam, and Raymond Chiong. "Deep learning for human affect recognition: Insights and new developments." *IEEE Transactions on Affective Computing* (2019).

45. Jack Jr, Clifford R., Matt A. Bernstein, Nick C. Fox, Paul Thompson, Gene Alexander, Danielle Harvey, Bret Borowski, et al. "The Alzheimer's disease neuroimaging initiative (ADNI): MRI methods." *Journal of Magnetic Resonance Imaging* 27, no. 4 (2008): 685–691.

46. Ellis, Kathryn A., Ashley I. Bush, David Darby, Daniela De Fazio, Jonathan Foster, Peter Hudson, Nicola T. Lautenschlager, et al. "The Australian Imaging, Biomarkers and Lifestyle (AIBL) study of aging: Methodology and baseline characteristics of 1112 individuals recruited for a longitudinal study of Alzheimer's disease." *International Psychogeriatrics* 21, no. 4 (2009): 672–687.

47. Marcus, Daniel S., Tracy H. Wang, Jamie Parker, John G. Csernansky, John C. Morris, and Randy L. Buckner. "Open Access Series of Imaging Studies (OASIS): cross-sectional MRI data in young, middle aged, nondemented, and demented older adults." *Journal of Cognitive Neuroscience* 19, no. 9 (2007): 1498–1507.

48. Malone, Ian B., David Cash, Gerard R. Ridgway, David G. MacManus, Sebastien Ourselin, Nick C. Fox, and Jonathan M. Schott. "MIRIAD—Public release of a multiple time point Alzheimer's MR imaging dataset." *NeuroImage* 70 (2013): 33–36

12 A Text Analytics-based E-Healthcare Decision Support Model Using Machine Learning Techniques

A. Sheik Abdullah, R. Parkavi, P. Karthikeyan, and S. Selvakumar

CONTENTS

12.1 INTRODUCTION TO DATA ANALYTICS

The domain of data analytics corresponds to the study of raw data with the mechanism of concluding solutions for a real-time problem analysis. In recent days, the term analytics has been widely used by most of the educational and industrial organizations for decisions-making at various levels. The level of forming solutions and interpretations can be made easier with the analytic approaches and its algorithmic patterns. The extraction of patterns and decision making differs significantly from data mining by the use of algorithm, model development, interpretation, and evaluation.

12.2 ANALYTICS PROCESS MODEL

Analytics uses different sorts of algorithms, including machine learning, knowledge discovery, and pattern analysis. The rate of modeling and its analysis factors are the cases that have to be considered for interpretation and evaluation. The raw data may contain various sorts of missing fields, data duplication, and varying formats. This has to be identified in the first stage of analysis.

The second stage works to transform the data and make it available for the modeling process. All modeling mechanisms don't always fit for the selected data—some may work well and some may lead to poor performance analysis. This has to be consistently checked with the parameters that correspond to the algorithm and its evaluation. The final stage is data interpretation and evaluation. Figure 12.1 provides an overview about the analytics process model.

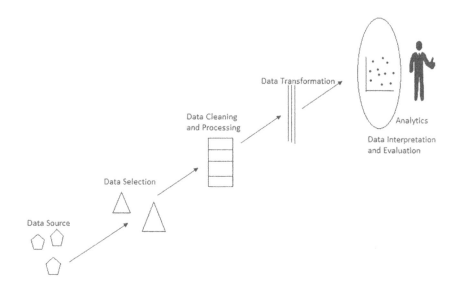

FIGURE 12.1 Analytics process and its stages.

12.3   REQUIREMENTS OF AN ANALYTICAL MODEL

The analytic model must have the capability to solve the problem for which it has been developed. The model should adhere to the nature of predictive capabilities with clear interpretability and justifiability. It should have a good statistical interpretation upon proof validation. The efficiency should be related to the data analysis process, its capabilities, and legal practices and standards. In general, there are five different types of analytics:

1. Predictive analytics
2. Text analytics
3. Descriptive analytics
4. Survival analytics
5. Social media analytics

12.4   TEXT ANALYTICS PROCESS MODEL

The process of text analytics is also referred to as text mining or opinion mining. It usually involves the analysis of extracting the necessary information from documents. The process involves natural language processing (NLP) with suitable opinion extraction information retrieval (IR) methods. It includes various subprocesses relative to that of the task involved.

In information retrieval, a vector-space model is considered to be the best model to provide answers, which people find similar to a search query. Usually, weighted vectors are used to represent the documents in the vector space model and provide the document that is more relevant to the user query. In the vector-space model, the similarities of a document to the query are computed by the weights provided to the term, called index terms, are assigned to each document's terms. There are many other ways to compute the weight of a document and the challenge we continuously face is in finding the best weight terms. The most-used schemes here are calculating the times the index term occurs in the document and calculating the index terms' occurrence in the total number of documents. The main terms used are term frequency and inverse term frequency. Upon considering medical data, the process of stemming, stop word removal, and similar preprocessing stages play a significant role in data preparation and analysis.

The process of sentiment analysis is:

• Document stage
• Sentence stage
• Aspect level of analysis

The mechanism of aspect level includes named entity and feature level modeling.

12.5   ONTOLOGY IN TEXT MINING WITH MEDICAL DATA

Currently, medical text data is growing rapidly and it includes patient details with corresponding clinical data. Maintaining medical record is a tedious job due to its terminology and complexity. In this chapter, the authors used a concept graph for medical document classification using Unified Medical Language System (UMLS). Datasets used in this chapter are: the Consumed dataset and the Medical Notes dataset. A medical text document is given as input to the UMLS, it finds the list of medical concepts with supervised concept weighting, a concept graph is constructed with the set of weights, given weights for edge and node, and finally, an enriched concept graph is generated for the given medical text document. The proposed method is measured with recall and F1-scores. When compared with other methods like Class Meaning Kernel and Class Weighting Kernel, the proposed method performs well [1].

Clinical Ontology Based Information Extraction (COBIE) is proposed with the aim of providing a clear system for making clinical decisions. The proposed framework includes three phases. In the first phase, medical reports are given as the input data, medical data is extracted using information extraction techniques that generate and modify the ontology based on the evaluation process. The modified ontology is stored in the ontology knowledge store, and finally, a source ontology is created. The second phase includes the medical reports that are extracted using information extraction techniques; the extracted data is stored as target ontology. In the third phase, the action is to map the source and the level of ontology in order to produce mapping ontology. Mapping ontology is presented using color-coding visualization techniques, it helps the clinicians to understand and make decisions [2].

In this chapter, the authors present a machine learning model for identifying clinical terms from unstructured medical data with the use of ontology. The Neural Concept Recognizer model is proposed and it employs a convolution neural network. The proposed model is trained with two ontologies and the authors conclude that the proposed model produces high accuracy with F1 scores [3].

In this chapter, the authors propose the whale optimization with support vector machine (WO-SVM) method for classifying medical data using ontology. In the proposed approach for privacy preservation, a Kronecker bat algorithm is deployed. The evaluation metrics used in the proposed method are accuracy, sensitivity, specificity, and fitness. The proposed method WO-SVM is compared with other methods such Naive Bayes, KNN, decision tree, and SVM. The authors conclude that WO-SVM has highest evaluation metrics compared to the other methods [4].

12.6 TEXT ANALYTICS APPLICATIONS

Text analytics work with huge amounts of unstructured data and it grows exponentially in terms of relevance and quantity. Like the other techniques i.e., text categorization, analytics, extraction of hidden information, and Region of Interest (ROI). Text analytics are used in many applications; the following applications are the most important:

1. Risk management
2. Knowledge management
3. Cybercrime prevention
4. Customer care service
5. Fraud detection through claims investigation

6. Contextual Advertising
7. Business intelligence
8. Content enrichment
9. Spam filtering
10. Social media data analysis

The authors propose a systematic review model that includes four techniques of text analytics, finding the strength and weakness of text analytics methods at each stage in the systematic process. The authors conclude that systematic reviews are broadly established to investigate research [5].

The authors use text analytics and content analysis techniques, identified in four major areas, such as psycho-pathology, patient perspective, medical records, and medical literature. Text mining techniques are used to extract information from complex and unstructured data in psychiatry [6].

People are using "smart" services everywhere from home to office. The authors use a data-driven approach for understanding smart services by using text mining and other machine learning algorithms and provide various research topics related to multiple smart service systems, such as smart cards, antennas, homes, phones, health, citites, energy management, grids, sensing, Internet of Things (IoT), etc. [7].

Financial forecasting applications include stock market prediction. Foreign Exchange (FOREX) automates human-like thinking and helps make decisions in financial organizations, thereby reducing financial risk. Text mining in financial applications includes three phases: financial forecasting, customer relationship management, and cyber security. Various studies are conducted in the field of FOREX market rate prediction and detection. To predict the future market rate, the investor must thoroughly study the historical data and also access the current situations of the market rate. Data are collected from financial news, social media, news headlines, tweets, and financial annual reports [8].

12.7 ELECTRONIC DATA PROCESS

Medical organizations use electronic medical records (EMR) to maintain patient medical reports; it includes details such as the patient health condition, diagnostic information, and treatment information. Even though it is maintained electronically, due to the EMR characteristics, it is difficult to extract the data. Most of the medical data are unstructured data; for extracting medical information, named-entity recognition and relation extraction are used. EMR data is split into three categories: structured, semi-structured, and unstructured data. Data preprocessing of EMR includes cleaning, integration, transformation, reduction, and privacy preservation. The source data is converted into processed data after it completes the data preprocessing steps [9].

In this chapter, the authors describe a systematic NLP and text analytics and apply these two techniques to extract the symptoms and process the electronic patient-authored text (ePAT). They found that the most clinical symptoms are pain, fatigue, and disturbance in sleep [10].

Authors used a clinical data collector tool to collect both the structured and unstructured electronic health data. After preprocessing and NLP, electronic health data (EHD) is stored in the SQL database. Next, the data are moved to the search engine and then to the client's machine where the user builds the queries. Finally, data are moved to the UMLS to perform the text mining process. The authors concluded that efficiency is improved by using the clinical data collector with real-world data [11].

Data are collected based on the information provided by the patient during admission in the hospital and includes general information about the patient, radiology related questions, radiology reports, and pathology reports. Various diseases considered in this study are: lung cancer, breast cancer, colon cancer, etc. SVM is used to perform classification and metrics used for evaluation are precision, recall and F-score. The classification percentage is improved with the real dataset and no single method can be applied to all diseases [12].

Privacy-preserving Gaussian distributed independent frequently subsequence algorithm (ppGDIFSEA) is proposed to assure the privacy preservation in text mining medical data for COVID-19 cases. The proposed method ppGDIFSEA employs recurrent neural network deep learning algorithm. Data are collected from medical records, papers, and pathological statistics and are then recognized with an optical character reader. Data classification and privacy preservation with ppGDIFSEA ensures it accurately and effectively found the key patterns [13].

12.8   ETHICS

There are three major areas of text analytics related to mental health: experimental settings, social media, and Electronic Health Records EHR. Using online social media data without getting proper permission from the user is unethical. The authors describe recent research work in text mining and mental health, current ethical issues, and also provide suggestions to overcome ethical issues [14].

12.9   TELEMEDICINE AND CONSULTATION PROCESS

The process of telemedicine is mainly concerned with the critical factor with Information and Communication Technology (ICT) to transfer and exchange valid information for treatment analysis with an improvement toward community-based health service. Meanwhile, telehealth is concerned with assisting health-related services through means of telecommunication and digital technologies. In India, healthcare monitoring and analysis is a challenging and most needed with newly originating diseases and syndromes. In this concern, travel and cost play a significant role in clinical data prediction and analysis. Patients with fewer syndromes at certain stages have less need to travel for doctors and therapist consultations. They can use a telemedicine facility for cost savings and treatment analysis with better exposure for getting well [15].

Telemedicine at all stages can render optimal service with faster and efficient treatment analysis. At certain stages, the patient is unable to get a medical consultation and service due to family or other societal impacts. In these cases, telemedicine will play a significant role for the patients to receive consultation from doctors and healthcare providers. In case of routine check-ups and regular monitoring, telemedicine can be used to ensure patient safety and well-being conditions. Upon adhering to the telemedicine processs, there is a higher likelihood for managing the patient's records and thereby minimizing the likelihood of patients missing advice from doctors and healthcare providers. It also enhances patient safety and precautionary measures, which can be incorporated into the technology [16].

12.10 E-HEALTHCARE DECISION SUPPORT MODEL

Health care data analytics include the collection of healthcare data (patient behavior, clinical results, claims, and costs) to formulate a decision plan that involves the discovery, interpretation, and exploration of meaningful patterns in the data. With the rapid development in tools and procedures over healthcare, it is necessary that healthcare data need to be analyzed for decision-making at proper intervals [17]. Clinical prediction models are most needed in the domain of healthcare technology. Different models have been used in practice for predicting different levels of syndromes. Most all of the models deploy predictive mechanisms with statistical significance. The outcome of the model provides the estimation of medical cost, region-based analysis of attribute determination, significance among the features, and the correlation patterns that exist among the attributes [18]. Meanwhile, the type of the diagnostic problem with the risk of patient's behavior can also be analyzed with the prediction model [19]. The following are the algorithms that have been used for the deployment of the decision support model in healthcare analytics.

12.10.1 Decision Trees

In 1986, John Ross Quinlan, a research scholar in the domain of machine-learning, developed Iterative Dichotomiser 3 (ID3), a decision tree algorithm. In accordance with this algorithm on ID3, Quinlan proposed a successor of ID3 algorithm known to be the C4.5 algorithm. This algorithm became more familiar with the data classification techniques corresponding to supervised learning systems. The algorithm ID3 is suitable for decision criteria with smaller samples of categorical data. The algorithm follows a greedy approach in the art of determining the local optimal value. Over-fitting of values may occur at stages where the algorithm fails to produce the optimal result on the given data [20].

12.10.2 Artificial Neural Network

The functionality of artificial neural network is stimulated by biological neural systems in which nodes called "neurons" have been combined to formulate a weighted link to form network of neurons. The neuron is said to be the computing component, which has adaptive weights that then produce the output based upon the activation function [21]. The perceptron neural function has only its input and output layers. The goal of this method is to solve problems in accordance to the working of human brain function.

12.10.3 Kernel Methods

The principle behind kernel methods is that the attributes in the given search space are mapped to an abstract space that provides a way to differentiate among multiclass classifications. The performance of kernel methods affords good results with data projection over large dimensions. The challenge behind kernel methods lies in choosing the right kernel function for the right data over evaluation [22]. Among the given data objects, the kernel function evaluates the similarity with the assignment in the highest value for kernel as $K(X,X')$.

12.10.4 Cost-Sensitive Methods

Clinical prediction models are typically developed with the aim to determine misclassification and test costs. Among all the various cost determination strategies, misclassification and test cost that eventually determine the efficacy of the computational algorithms. The model design in clinical prediction relies upon the computational cost and instability of the algorithms, which need to be addressed with each dataset.

12.11 ROLE OF MACHINE LEARNING IN HEALTH CARE

A model for predicting healthcare of a patient through an apparatus was proposed by Steven Johnson and Christopher Busch in 2011 [11]. Recently, robots are working along with healthcare teams for decision making [23]. The predominant method in this system is a machine learning (ML) algorithm [24]. ML with text classification works in real-time surveillance in the medical domain, particularly in health informatics [25]. Automated ML is another major trend with many possible applications in the scope of healthcare research [26]. Prediction of patient healthcare consumption using ML is another approach [27]. In cardiovascular disease, diagnosis, and monitoring, ML with network analytics has been proposed with existing type 2 diabetes patients [28].

12.12 CASE STUDY—TOOL DEMONSTRATION

12.12.1 Case Study: Medical Fraud Detection

The following case study discusses medical fraud detection and its experimental analysis.

The aim is to detect medical fraud based on patient information by training and applying the gradient boosted trees (GBT) model.

Step 1: Collect the medical data from patients with the historical information and its corresponding fraudulent actions. The model is set using GBT algorithm.
Step 2: Some of the attributes in the dataset seem to be correlated. Higher correlation is removed with a significance rate of 95%.
Step 3: The algorithm is specifically used for fraudulent monitoring and analysis. The balancing of data is made to detect the fraudulent actions.

Parameter settings:
Number of folds = 10

TABLE 12.1

Experimental Results

	True False	True True	Class Precision
Pred. false	69	5	93.24%
Pred. true	31	95	75.40%
Class recall	69.00%	95.00%	

Sampling type = automatic

Significant experimental results were observed with the generation of 20 trees in the platform and the most predominant attribute—amount paid to date—is found to be most important. Among the 16 attributes, it was found that the most forms of intrusion happened across the amount paid attribute out of all the attributes in the tuple of records. Table 12.1 provides the experimental results as observed for the data

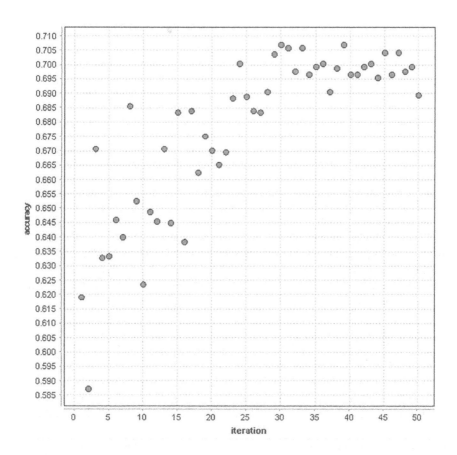

FIGURE 12.2   Accuracy estimation iteration vs. accuracy.

TABLE 12.2
Sample Prediction Data with Confidence Level

Machine_ID	Confidence(no)	Confidence(yes)	Prediction(Failure)
M_0221	0.1	0.9	yes
M_0223	0.1	0.9	yes
M_0271	0.2	0.8	yes
M_0151	0.2	0.8	yes
M_0167	0.2	0.8	yes
M_0176	0.2	0.8	yes
M_0177	0.2	0.8	yes
M_0190	0.2	0.8	yes
M_0192	0.2	0.8	yes
M_0225	0.2	0.8	yes
M_0227	0.2	0.8	yes
M_0239	0.2	0.8	yes
M_0260	0.2	0.8	yes
M_0262	0.2	0.8	yes
M_0137	0.2	0.8	yes
M_0140	0.2	0.8	yes
M_0153	0.2	0.8	yes
M_0154	0.2	0.8	yes
M_0162	0.2	0.8	yes
M_0166	0.2	0.8	yes
M_0194	0.2	0.8	yes
M_0212	0.2	0.8	yes
M_0219	0.2	0.8	yes

corresponding to fraudulent actions. The estimated iterations to that of accuracy can be viewed at Figure 12.2.

Hence, from the experimental results it has been observed that accuracy has been met with an improvement of about 82% for the dataset considered. Thus, from these significant algorithms, we can also estimate the fraudulent actions concerned in data access and manipulation. The Table 12.2 provides the sample prediction to that of the confidence value.

12.12.2   CASE STUDY: PREDICTIVE MAINTENANCE

Step 1: load the data corresponding to the simulation.
Step 2: determine the most influencing factors by average weighting
Step 3: train a KNN model optimizing for the K to ensure maximum accuracy
Step 4: load the new data and apply the corresponding model to predict failure

Predictive maintenance and fraud detection analysis are some of the real-time case studies used for the predictive analysis of textual data in the healthcare sector. This

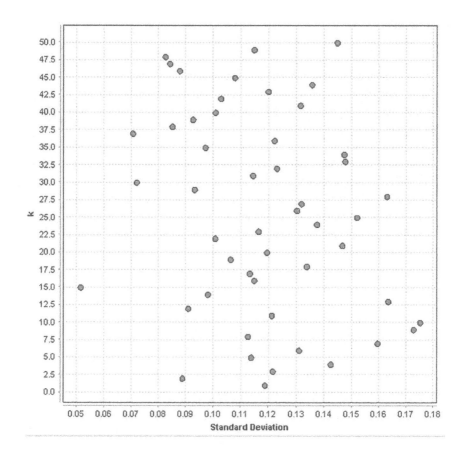

FIGURE 12.3 Standard deviation estimation SD vs. K value.

process if deployed in medical and clinical domains can detect more fraud and error rates to the most significant levels. The estimated SD to that of the k value can be viewed at Figure 12.3

12.13 CONCLUSION

This chapter focuses on the approaches and paradigms that correspond to the medical data analysis in a textual perspective. ML algorithms corresponding to decision trees, neural network, kernel methods, random forest, and boosting techniques are the most used data models for predicting the nature of textual data in the medical domain. Text processing and analysis involves different feature analysis steps, including preprocessing, categorization, sentiment evaluation, and interpretation. The utilization of the right data in the right platform can bring significant pattern analysis and evaluation. This should be kept in mind for any sort of data processing models when dealing with text methods and the utilization of ML algorithms— valuable service can be rendered in the right place at the right time.

REFERENCES

1. Karlekar, Nandkishor P. and N Gomathi. "OW-SVM: Ontology and whale optimiza-tionbased support vector machine for privacy-preserved medical data classification in cloud," *International Journal of Communication System*, 31, no. 9 (2018), e 3700. doi: https://doi.org/10.1002/dac.3700

2. Waring, Jonathan, Charlotta Lindvall, Renato Umeton. "Automated machine learning: Review of the state-of-the-art and opportunities for healthcare," *Artificial Intelligence in Medicine*, 104 (2020).

3. Beckman, Adam L., Julie A. Shah, Neel T. Shah, Sanchay Gupta. "Robots join the care team: Making healthcare decisions safer with machine learning and robotics," *Healthcare*, 8, no. 4 (2020).

4. Mueller-Peltzer, M., Feuerriegel, S., Molgaard Nielsen, A., Kongsted, A., Vach, W., and Neumann, D. "Longitudinal healthcare analytics for disease management: Empirical demonstration for low back pain," *Decision Support Systems*, 132 (2020), 113271. doi:10.1016/j.dss.2020.113271

5. van Laar, Sylvia A., Kim B. Gombert-Handoko, Henk-Jan Guchelaar and Juliëtte Zwaveling. "An electronic health record text mining tool to collect real-world drug treatment outcomes: A validation study in patients with metastatic renal cell carci-noma," *Clinical Pharmacology and Therapeutics*, 18, no. 3 (2018). doi: 10.1002.cpt. 1966.

6. Abbe, Adeline, Cyril Grouin, Pierre Zweigenbaum, and Bruno Falissard. "Text min-ing applications in psychiatry: a systematic literature review." *International Journal of Methods in Psychiatric Research*, 25, no. 2 (2015), 86–100. doi:10.1002/mpr.1481

7. Dreisbach, Caitlin, Theresa A. Koleck, Philip E. Bourne, and Suzanne Bakken. "A systematic review of natural language processing and text mining of symptoms from electronic patient-authored text data," *International Journal of Medical Informatics*, 125 (2019), 37–46. doi:10.1016/j.ijmedinf.2019.02.008.

8. Abdullah, A. Sheik and R. R. Rajalaxmi. "A data mining model for predicting the coronary heart disease using random forest classifier," *IJCA*, 3 (2012), 22–25. Available from: https://www.ijcaonline.org/proceedings/icon3c/number3/6020-1021

9. Thomas, James, John McNaught, and Sophia Ananiadou. "Applications of text min-ing within systematic reviews," Research Synthesis Methods, 2, no. 1 (2011), 1–14. doi:10.1002/jrsm.27.

10. Ma, Bo, Jinsong Wu, Shuang Song, William Liu. "Assuring privacy-preservation in mining medical text materials for COVID-19 cases—A natural language perspec-tive," *Open Journal of Internet of Things,* 6, no. 1 (2020). http://www.ronpub.com/ojiot ISSN 2364-7108

11. Johnson, Steven and Christopher Busch. "Method and apparatus for assessing credit for healthcare patients," US-Patent, 2011.

12. Kumar B. Shravan and Vadlamani Ravi. "A survey of the applications of text mining in financial domain," *Knowledge-Based Systems* 114 (2016), 128–147. doi: https://doi.org/10.1016/j.knosys.2016.10.003

13. Arbabi, Aryan, David R. Adams, Sania Fidler, Michael Brudno. "Identifying clinical terms in medical text using ontology-guided machine learning," *JMIR Med Inform*, 7, no. 2 (2019):e12596. doi: 10.2196/12596

14. Skorburg Joshua August and Phoebe Friesen, "Ethical issues in text mining for mental health," Penultimate Draft (June 2020), Forthcoming in M. Dehghani & R. Boyd (Eds.) *The Atlas of Language Analysis in Psychology*. Guilford Press.

15. KajaNisha R. and A. Sheik Abdullah, "Classification of cancer micro-array data with feature selection using swarm intelligence techniques," *Acta Scientific Medical Sciences*, 3, no. 7 (2019), 82–87.

16. McLeod, A. and D. Dolezel. "Cyber-analytics: Modeling factors associated with healthcare data breaches." *Decision Support Systems*, 108 (2018), 57–68. doi:10.1016/j.dss.2018.02.007.

17. Abdullah, A. Sheik, S. Selvakumar, P. Karthikeyan, M. Venkatesh. "Comparing the efficacy of decision tree and its variants using medical data," *Indian Journal of Science and Technology*, 10 no. 18 (2017). doi: 10.17485/ijst/2017/v10i18/111768

18. Sheik Abdullah, A, Akash, K, Shamin Thres, J, Selvakumar, S. "Sentiment Analysis of Movie Reviews using Support Vector Machine Classifier with Linear Kernel Function, Frontiers in Intelligent Computing: Theory and Applications (FICTA 2020), Volume 1, Evolution in Computational Intelligence, Springer, ISBN 978-981-15-5787-3. https://doi.org/10.1007/978-981-15-5788-0_34

19. Ekramul Hossain Md, Shahadat Uddin, Arif Khan. "Network analytics and machine learning for predictive risk modelling of cardiovascular disease in patients with type 2 diabetes," *Expert Systems with Applications*, 164 (2021).

20. Abdullah, A. Sheik and P. Priyadharshini. "Big Data and Analytics," for the book Big Data Analytics for Sustainable Computing. IGI Global Publishers (2019), pp. 47–56. doi: 10.4018/978-1-5225-9750-6.ch003.

21. Abdullah, A. Sheik, "A data mining model to predict and analyze the events related to coronary heart disease using decision trees with particle swarm optimization for feature selection," *International Journal of Computer Applications*, 55, no. 8 (2012), pp. 49–55.

22. Selvakumar S., A. Sheik Abdullah, R. Suganya. "Decision support system for type II diabetes and its risk factor prediction using bee based harmony search and decision tree algorithm," *International Journal of Biomedical Engineering and Technology*, 29, no. 1 (2019).

23. Kocbek, Simon, Lawrence Cavedon, David Martinez, Christopher Bain, Chris Mac Manus, Gholamreza Haffari, Ingrid Zukerman, Karin Verspoor. "Text mining electronic hospital records to automatically classify admissions against disease: Measuring the impact of linking data sources," *Journal of Biomedical Informatics*, 64 (2016), 158–167. doi: 10. https://doi.org/10.1016/j.jbi.2016.10.008.

24. Akash K., Sheik Abdullah A. (2021) A New Model of Zero Energy Air Cooler: An Cost and Energy Efficient Device in Exploit. In: Kumar J., Jena P. (eds) Recent Advances in Power Electronics and Drives. Lecture Notes in Electrical Engineering, vol 707. Springer, Singapore. https://doi.org/10.1007/978-981-15-8586-9_14

25. Gupta, Aakansha and Rahul Katarya. "Social media based surveillance systems for healthcare using machine learning: A systematic review," *Journal of Biomedical Informatics*, 108 (2020).

26. Harsheni, S. K, S. Souganthika, K. GokulKarthik, A. Sheik Abdullah, S. Selvakumar, "Analysis of the Risk factors of Heart disease using Step-wise Regression with Statistical evaluation," International Conference on Emerging Current Trends in Computing and Expert Technology, (COMET 2019), (ISSN: 2367-4512). March 2019, 22–23.

27. Shanavas, Niloofer, Hui Wang, Zhiwei Lin, Glenn Hawe. "Ontology-based enriched concept graphs for medical document classification," *Information Sciences*, 525 (2020), 172–181.

28. Lim, Chiehyeon, and Paul P. Maglio. "Data-driven understanding of smart service systems through text mining." *Service Science* 10, no. 2 (2018), 154–180. doi: 10.1287/serv.2018.0208

Index

Printed in the USA
CPSIA information can be obtained
at www.ICGtesting.com
LVHW021733041124
795688LV00039B/1208